Hybrid Nanofluids

This book provides a comprehensive understanding of advanced hybrid nanofluid applications in various fields while also explaining the real-time industrial applications of nanofluids. It explains mathematical, numerical, and experimental methodologies of application of the nanofluids in heat transfer and mass transfer processes. It helps build innovative nanofluid-based devices, including the study and measurement of thermophysical characteristics, convection, and heat transfer equipment performance.

Features:

- Discusses hybrid nanofluids with a strong attention to the processes.
- Explores inter-relation between thermal properties, physical properties, and optical properties of the nanofluids.
- Investigates high-performance heat transfer and mass transfer hybrid nanofluids.
- Explores data for the design of the nanofluid application and scale-up challenges.
- Reviews industrial operation and scale-up challenges for nanofluid applications in the industrial process.

This book is aimed at graduate students and researchers in fluid dynamics, nanotechnology, and chemical and mechanical engineering.

Emerging Materials and Technologies

Series Editor: Boris I. Kharissov

The *Emerging Materials and Technologies* series is devoted to highlighting publications centered on emerging advanced materials and novel technologies. Attention is paid to those newly discovered or applied materials with potential to solve pressing societal problems and improve quality of life, corresponding to environmental protection, medicine, communications, energy, transportation, advanced manufacturing, and related areas.

The series takes into account that, under present strong demands for energy, material, and cost savings, as well as heavy contamination problems and worldwide pandemic conditions, the area of emerging materials and related scalable technologies is a highly interdisciplinary field, with the need for researchers, professionals, and academics across the spectrum of engineering and technological disciplines. The main objective of this book series is to attract more attention to these materials and technologies and invite conversation among the international R&D community.

Hydrogen Production, Storage, and Utilization
Technologies and Applications
Abbas Tcharkhtchi, Hamidreza Vanaei, Albert Lucas, and Sedigheh Farzaneh

MXenes for Energy Storage Applications
Emerging Characteristics, Compositions, and Synthesis Methods
Muhammad Rafique, M. Bilal Tahir, and Saira Anwar

Multi-scale and Multifunctional Coatings and Interfaces for Tribological Contacts
Ajit Behera, Kuldeep K Saxena, Dipen Kumar Rajak, and Shankar Sehgal

Nanotechnology in Green Energy Generation
Ahmed Thabet Mohamed

Thermo-Acoustics of Nanofluids and Transfer Processes
Shriram S. Sonawane and Manjakuppam Malika

Hybrid Nanofluids
Heat and Mass Transfer Processes
Shriram S. Sonawane and Manjakuppam Malika

Applications of Hybrid Nanofluids in Science and Engineering
Edited by A. K. Pandey, H. Upreti, O. D. Makinde, and A. J. Chamkha

For more information about this series, please visit: www.routledge.com/Emerging-Materials-and-Technologies/book-series/CRCEMT

Hybrid Nanofluids

Heat and Mass Transfer Processes

Shriram S. Sonawane and Manjakuppam Malika

CRC Press is an imprint of the
Taylor & Francis Group, an **informa** business

Designed cover image: shutterstock

First edition published 2025
by CRC Press
2385 NW Executive Center Drive, Suite 320, Boca Raton FL 33431

and by CRC Press
4 Park Square, Milton Park, Abingdon, Oxon, OX14 4RN

CRC Press is an imprint of Taylor & Francis Group, LLC

ISBN: 978-1-032-60795-5 (hbk)
ISBN: 978-1-032-97730-0 (pbk)
ISBN: 978-1-003-59513-7 (ebk)

DOI: 10.1201/9781003595137

Typeset in Times
by Apex CoVantage, LLC

Dr. SHRIRAM S. SONAWANE

I dedicate this book to my loving family—my mother "AAI", my father "DADA", my most loving wife Rina, my gorgeous daughter "Sanyuja", and my son "Shaurya". Your unwavering support and encouragement have been my guiding light throughout this incredible journey.

Dr. MANJAKUPPAM MALIKA

This work is dedicated to my parents, Mrs. Nirmala and Mr. Gopal, and my sisters, Mrs. Monica and Mrs. Harika, whose unwavering support and encouragement fostered my passion for science and technology.

I also extend my heartfelt gratitude to my beloved husband, Mr. Rahul Mudliar, and my daughter, Ms. Prisha Mudliar, for their patience and constant support throughout this journey. Special thanks to my in-laws, Mrs. Sujatha Mudliar and Mr. Narendra Mudliar, for their understanding and help during the course of this project.

Contents

Preface

Hybrid Nanofluids: Heat and Mass Transfer Processes provides a comprehensive understanding of advanced hybrid nanofluid applications in a variety of fields, while also explaining the real-time industrial applications of nanofluids.

- The book covers fundamental topics such as preparation studies, analysis, and measurement of thermophysical properties, as well as various industrial applications, providing a rigorous framework to assist readers in developing new nanofluid-based devices. It also investigates the environmental implications of nanofluid use using life cycle analysis and SWOT analysis.
- In addition to the existing books, the proposed book can be used by readers to thoroughly understand the fundamentals of nanofluids and to further study how to use them in any real-time applications.
- This book has no prerequisites and is intended for scientists/scholars in any field of science and technology. As a result, it includes all the fundamental information required for beginners. So, by reading the book, regardless of the background, the authors can learn how to create nanofluids, which can be used as coolants or in any new applications.
- It provides detailed knowledge of mathematical, numerical, and experimental methodologies of application of the nanofluids in heat transfer and mass transfer processes.
- This book provides a rigorous framework to aid readers in building innovative nanofluid-based devices, covering essential topics such as the study and measurement of thermophysical characteristics, convection, and heat transfer equipment performance.
- Various edited books are available in the market on the heat transfer application of nanofluids, but there is no authored book that incorporates the effect of both heat and mass transfer for the successful industrial application of nanofluids.
- This book is divided into three main sections:

  i. Fundamentals of nanofluids and hybrid nanofluids
  ii. Application of nanofluids for the heat and mass transfer processes
  iii. Process analysis and optimization

    - Advanced numerical approach
    - Life cycle analysis
    - Industrial scale-up challenges
    - Future works

The first section investigates the underlying mechanisms of thermal transport in nanofluids using a combination of theoretical, experimental, and numerical research. It gives a detailed overview of the fundamentals of nanofluids and the preparation and characterization of hybrid nanofluids. The second section gives a detailed overview of the application of nanofluids in heat and mass transfer operations. The third section includes a complete SWOT analysis based on optimization techniques for the mass transfer and heat transfer operations, socio-economic and environmental impacts of nanofluids using LCA, and industrial operation and scale-up challenges for nanofluid applications in the industrial process.

This book is useful for researchers and professionals who are working in industry/academia or anyone interested in the applications of nanofluids in industrial processes for design purposes. The following sectors of the audience are the primary audience of the proposal.

1. UG/PG/doctoral students in the field of nanotechnology/chemical engineering and aspiring researchers within a broad domain of nanotechnology/chemical engineering
2. Faculty and staff from reputed academic institutions and technical institutions
3. Executive/engineer or researcher from the manufacturing and service industry of the chemical and biomedical industry
4. Government entities, such as R&D facilities, that have specialized research departments focused on nanotechnology and chemical processes
5. Professionals in the fields of chemical, materials, and mechanical engineering; nanoscience; soft matter physics; and chemistry

Foreword

The field of thermal management is undergoing significant transformation, driven by advances in nanotechnology and its application in heat and mass transfer processes. *Hybrid Nanofluids: Heat and Mass Transfer Processes* marks a major milestone, emphasizing the integration of nanofluids and hybridization to boost efficiency and functionality in thermal applications.

This book offers an in-depth exploration of nanofluids and hybrid nanofluids, from their origins and preparation methods to the factors influencing their properties. It addresses critical topics like stability enhancement techniques, covering both mechanical and chemical approaches to ensure sustained performance in real-world applications. The thermophysical, optical, electrical, magnetic, and dielectric properties of hybrid nanofluids are extensively discussed, along with valuable insights into nanoparticle mixing ratios and the enhancement of heat transfer mechanisms.

Highlighting the groundbreaking role of hybrid nanofluids in heat exchangers, the book also explores their diverse applications in wastewater treatment, membrane technology, and petroleum science, showcasing their adaptability across multiple industries. It concludes with a comprehensive analysis of the latest advancements, challenges, and future directions, including SWOT analysis and life cycle assessments, guiding readers through the evolving landscape of hybrid nanofluids.

As a critical reference for researchers, engineers, and professionals in chemical engineering, materials science, and environmental technology, *Hybrid Nanofluids: Heat and Mass Transfer Processes* presents the latest research and innovations in the field. It serves as a roadmap for those looking to harness the transformative potential of hybrid nanofluids, offering novel solutions for enhanced energy efficiency and sustainability across a range of applications.

Prof. Dr. Kamal Kishore Pant
Director, Indian Institute of Technology Roorkee, and Professor of Chemical Engineering, IIT Delhi

Author Biographies

Dr. Shriram S. Sonawane Professor, Department of Chemical Engineering Visvesvaraya National Institute of Technology (VNIT), Nagpur, India

Dr. Shriram S. Sonawane is an accomplished professor in the Department of Chemical Engineering at the Visvesvaraya National Institute of Technology (VNIT) in Nagpur, India. With over 22 years of experience in teaching and research, he has made significant contributions to the fields of chemical engineering. His research interests encompass a broad range of topics, including polymer nanocomposites, nanofluids, nano-separation technologies, process modeling and simulation, and extraction processes.

Dr. Sonawane has edited more than four books and is widely recognized for his expertise. He has been awarded the prestigious Fellow award by the Maharashtra Academy of Science and holds more than 12 Indian patents. To date, he has published over 200 research papers in renowned national and international journals, which have collectively garnered more than 6,000 citations according to Google Scholar. His academic influence extends to more than 100 presentations at national and international conferences, where he has also organized numerous symposia.

In his research, Dr. Sonawane has developed high-performance nanofluids for a variety of applications, including pool boiling, heat exchange, solar energy systems, wastewater treatment, petroleum technologies, membrane processes, and automotive cooling systems. Additionally, he and his team at VNIT have pioneered methods to enhance bio-hydrogen production from complex waste streams such as distillery wastewater, food waste, and dairy by-products. He has also innovated novel packing materials for use in extraction operations.

Dr. Sonawane is an active reviewer for several prestigious academic journals and serves as a Fellow of multiple professional organizations. He is also an esteemed member of various academic and governmental committees, contributing to policy and research development at both the national and institutional levels.

Manjakuppam Malika is an assistant professor in the Department of Chemical Engineering, BV Raju Institute of Technology (BVRIT), Narsapur, Telangana, India, She was recognized in Stanford University's "World's Top 2%" scientists list for the year of 2024. Her current research focuses on nanofluid applications in chemical engineering and technology.

1 Introduction to Hybrid Nanofluids

Efficient cooling is a crucial requirement for most industrial technologies. However, the inadequate thermal conductivity of heat transfer fluids poses a substantial obstacle that must be overcome to promote the progress of effective heat transfer coolants. The most recent developments in technology have given rise to the possibility of fabricating metallic or non-metallic particles with dimensions ranging within the nanoscale range. Nanomaterials have evolved as a remarkable category of materials encompassing a diverse variety of instances, characterized by at least one dimension falling under the 1 to 100 nanometers (nm) range,[1] where 1 nm is 10^{-9} m, which is 1,000 times smaller than the micron (μm) size (**Figure 1.1**). As the particle size diminishes, the surface-to-volume ratio of the particle is augmented, thereby resulting in a greater overall surface area for a given volume of material. Irrespective of the size, bulk material exhibits consistent physical–chemical characteristics. The sensible design of nanoparticles enables the attainment of remarkably elevated surface areas due to the increased percentage of atoms on their surface. With the increased surface energy, nanomaterials possess remarkable characteristics that significantly diverge from those shown by their bulk counterparts.

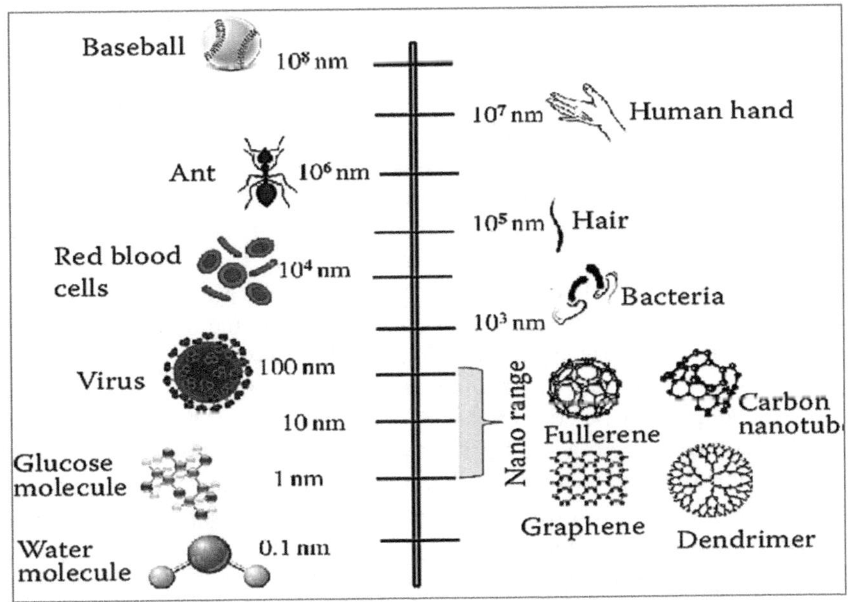

FIGURE 1.1 Comparison of nanoscale materials to well-known substances.[2]

DOI: 10.1201/9781003595137-1

Moreover, the properties of nanomaterials can be adjusted according to specific requirements by accurately manipulating factors such as size, shape, synthesis circumstances, and suitable functionalization techniques. Research has demonstrated that, within the nanometer scale, the distribution of particle sizes, the shape of particles, interfacial processes, and quantum effects significantly influence the modification of material properties. Nanotechnology has garnered attention due to its enhanced and modifiable characteristics, positioning it as a promising avenue for reshaping our viewpoints and aspirations while equipping us with the capacity to address global challenges. In recent times, the field of nanotechnology has grown as a multidisciplinary domain, whereby acquiring a fundamental comprehension of the thermal, physical, chemical, optical, and magnetic characteristics of nanostructures holds the potential to yield the subsequent production of useful materials with diverse uses.

1.1 ORIGIN OF NANOFLUIDS

Fluids are frequently utilized as heat transfer carriers in thermal energy exchange processes, for either heating or cooling purposes. However, as a result of the increased intermolecular spacing, fluids exhibit inferior efficiency in terms of heat transfer compared to solids. At standard ambient conditions, the majority of metals exhibit a substantial abundance of unbound electrons, hence facilitating the transfer of thermal energy to a greater extent in comparison to alternative substances. Based on the findings shown in **Table 1.1**, it is evident that the thermal conductivity of aluminum is approximately 400 times greater than that of water and 1,000 times greater than that of ethylene glycol. Metallic liquids exhibit higher thermal conductivity compared to non-metallic liquids. So, the inclusion of solid particles with enhanced thermal conductivity has the potential to significantly enhance the thermophysical characteristics of conventional fluids.

Prominent researchers have focused considerable work on improving the thermophysical properties of ordinary fluids due to the limitations imposed by thermal conductivity. In 1881, Maxwell conducted experiments including the inclusion of solid particles within a size range spanning from millimeters (mm) to micrometers (μm). These particles, possessing higher thermal conductivity, demonstrated a more favorable relative thermal conductivity than the primary fluid. Maxwell derived an equation to characterize the relative thermal conductivity of a fluid in the presence of spherical nanoparticles. This equation is contingent upon the concentration of particles. The increase in particle concentrations resulted in a higher aspect ratio of the particles, leading to the maximization of the effective heat conductivity. One of the primary challenges associated with suspensions that contain particles in the millimeter or micrometer size range is the issue of rapid sedimentation of these particles. In the event that the fluid is continuously circulated to minimize sedimentation, it is likely that the microparticles would cause erosion to the pipe walls, resulting in their gradual thinning. The experimental findings indicate that the pumping of suspensions containing micron-size particles can lead to significant clogging issues, rendering industrialization unfeasible. Following this, there weren't many significant advances in improving the thermophysical properties of heat transfer fluids.

TABLE 1.1
Thermal Conductivity (W/mK) of Different Materials at 25°C Room Temperature

Material	Thermal Conductivity (W/mK)
Fluids	
Deionized water	0.603
Hexane	0.126
Glycerol	0.285
Engine oil	0.145
Ethanol	0.172
Toluene	0.133
Ethylene glycol	0.252
Metallic solids	
Copper	401
Aluminum	257
Silver	429
Gold	345
Nickel	158
Iron	80
Non-metallic solids	
Carbon nanotubes	2000 to 6000
Diamond	2000
Silica	148
SiC	270
Al_2O_3	40
CuO	76
Fe_2O_3	7
ZnO	29
TiO_2	25

After nearly a century, a significant amount of research conducted in the late 1990s and early 2000s has shown that the addition of small amounts of nanoparticles to conventional base fluids, such as water, results in a significant increase in the thermal conductivity of the resulting fluid. In early 1992, researchers Choi and Eastman[3] were working on a liquid nitrogen cooling process using a microchannel system. Even though heat transfer was increased with the reduced channel diameter, they faced a limitation of increased pressure drop. As pressure drop and pumping power were directly related, there was a need to reduce the pressure drop. In 1995, leveraging Maxwell's theory, Choi pioneered an approach to address heat transfer challenges by significantly reducing particle sizes from the micrometer to the nanometer scale and dispersing these nanoparticles into a base fluid. He termed the resulting suspensions "nanofluids," recognizing their potential for enhanced thermal properties due to the unique behavior of particles at the nanoscale.

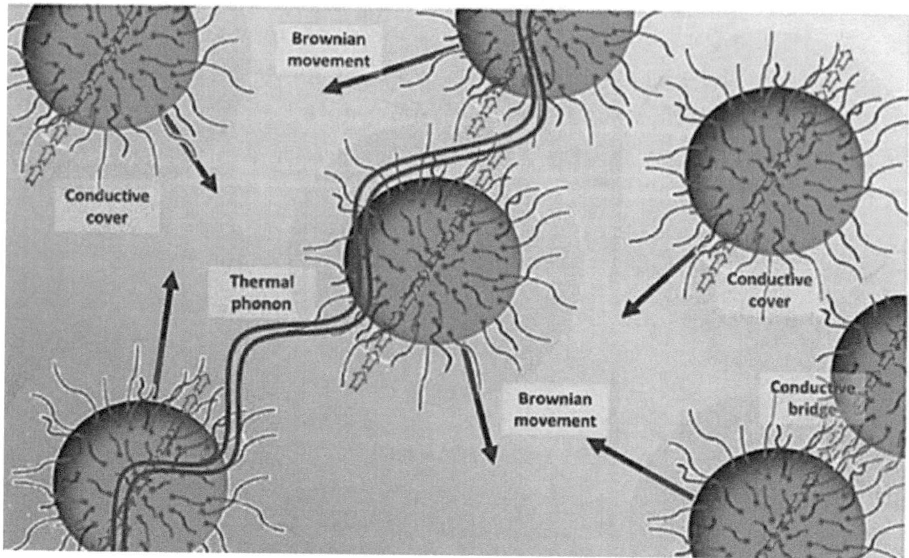

FIGURE 1.2 Thermal bridging phenomenon of nanofluids.[6]

Nanofluids refer to colloidal suspensions composed of nanoparticles, typically ranging in size from 1 to 100 nm dispersed inside liquid media known as base fluids. In a study conducted by Eastman et al. (2001),[4] a diluted 0.3 Vol% of copper nanoparticles (size: 10 nm) in ethylene glycol was utilized, leading to a 40% improvement in the conductivity of the resultant nanofluid. From then, several studies investigating mass transfer coefficients with nanofluids have generated significantly notable outcomes. It is anticipated that nanoparticles will be organized in a manner that promotes thermal bridging, hence augmenting the effective thermal conductivity of the underlying fluid as shown in **Figure 1.2**. The alignment of particles at the interface between nanoparticles and fluid facilitates the efficient passage of heat. Improvement can be achieved by the cumulative impact, as solids had superior thermophysical characteristics compared to fluids.[5] Furthermore, the particles inside the underlying fluid exhibit stochastic Brownian motion, hence resulting in frequent interactions among the particles and promoting the efficacy of the transfer mechanism.

1.2 ORIGIN OF HYBRID NANOFLUIDS

The efficacy of nanofluid is contingent on several factors, encompassing the nanoparticle type, shape, and concentration; the type of base fluid; the stability of the nanofluid; and the prevailing temperature circumstances throughout operation. The deliberate choice of a nanoparticle is a critical determinant in assessing the effectiveness of the resultant nanofluid, given that each nanoparticle exhibits unique characteristics. It is reasonable to anticipate that the incorporation of nanoparticles with higher thermal conductivity will lead to enhanced fluid performance. However, it should be noted that a majority of nanoparticles with elevated thermal conductivity,

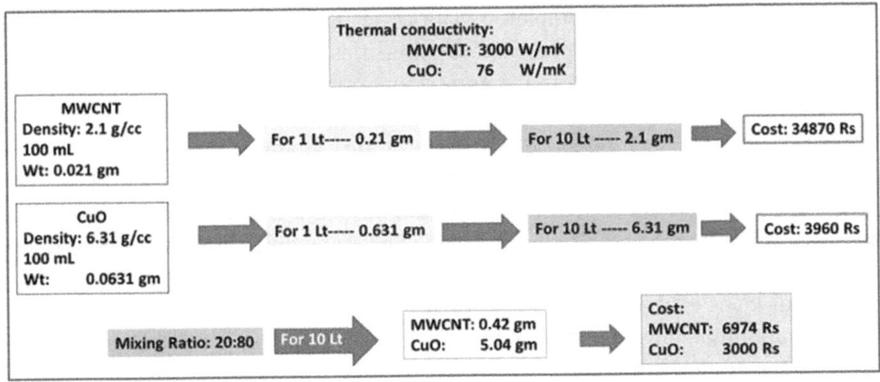

FIGURE 1.3 Cost estimation of nanofluid and hybrid nanofluid.

such as multi-walled carbon nanotubes (MWCNT), tend to have a greater cost than other types of nanoparticles. The cost of nanoparticles directly impacts the overall cost of the procedure. To reduce the overall costs associated with nanofluids, researchers have introduced a new class of fluids called hybrid nanofluids, which include the combination of several nanoparticles. It is anticipated that the incorporation of many nanoparticles will enhance the thermophysical properties in comparison to fluids based on a single nanoparticle. For example, as illustrated in **Figure 1.3**, if we were to produce 0.01 Vol% of 10 liters MWCNT/water nanofluid, we need to add roughly 2.1 gm of nanoparticle to water. Therefore, the cost of the 0.01 Vol% of 10 liters MWCNT/water nanofluid will be Rs. 34,870, and similarly for Rs. 3940 for CuO/water nanofluid. So researchers got a concept to employ both nanoparticles in different mixing ratios by which we can have the cumulative properties of both nanoparticles and also minimize the total cost of the nanofluid.

1.3 PREPARATION OF NANOFLUIDS

The synthesis of nanofluids encompasses several methodologies, which can be classified into two primary categories depending on the complexity of the procedure. There exist two primary methodologies for the preparation of nanofluids. That is to say, the approaches under consideration are one-step and two-step methods (**Figure 1.4**). Nanofluids can be synthesized using a one-step approach, wherein particular dispersion procedures are not required. This can be achieved through the utilization of techniques like as chemical vapor deposition, laser ablation, the submerged arc method, and similar methods. These methods enable the exact regulation of nanoparticle size and purity levels. The nanofluid generated from these technologies may exhibit enhanced stability, eliminating the need for further treatment methods such as magnetic stirring and ultrasonication for homogenization, pH adjustment, and surfactant incorporation. Nevertheless, the application of this approach is limited to a somewhat small scale due to the requirement of costly equipment setup. There are certain limitations associated with this strategy that should be taken into account. The persistence of residual reactants in nanofluids remains a notable problem due to incomplete reactions. This phenomenon is inevitable and cannot be circumvented. The presence of impurities cannot be disregarded, as they introduce a level of complexity to the assessment of nanoparticle impacts on a nanofluid. However, there have been reports documenting the synthesis of nanofluids by a simplified one-step technique. Munkhbayar et al.[7] employed a single-step evaporation technique to disperse silver nanoparticles onto nanofluids based on MWCNTs. Nanoparticles are produced via a process of concurrent evaporation and condensation under the influence of elevated voltage. The transmission electron microscopy analysis revealed that the silver nanoparticles exhibited the capability to adhere to the external layer of the MWCNTs in the absence of any surfactant. Moreover, the resultant nanofluid showed remarkable stability over a period of several days, exhibiting no signs of agglomeration.

The two-step approach involves the production of nanofluids by dispersing nanoparticles individually into the base fluid. In this process we have to fix the type of nanofluid we are preparing. During this process, it is essential to first determine the specific type of nanofluid being prepared. Based on this, the required amount of nanoparticles can then be calculated, as shown in **Equation 1.1**. The nanoparticles are introduced into the base fluid and subsequently homogenized through the utilization of mechanical stirring, ultrasonication, or a combination of both techniques. The formed nanofluids undergo additional treatments such as pH adjustment, surface alteration, or surfactant addition, depending on their dispersion stability.

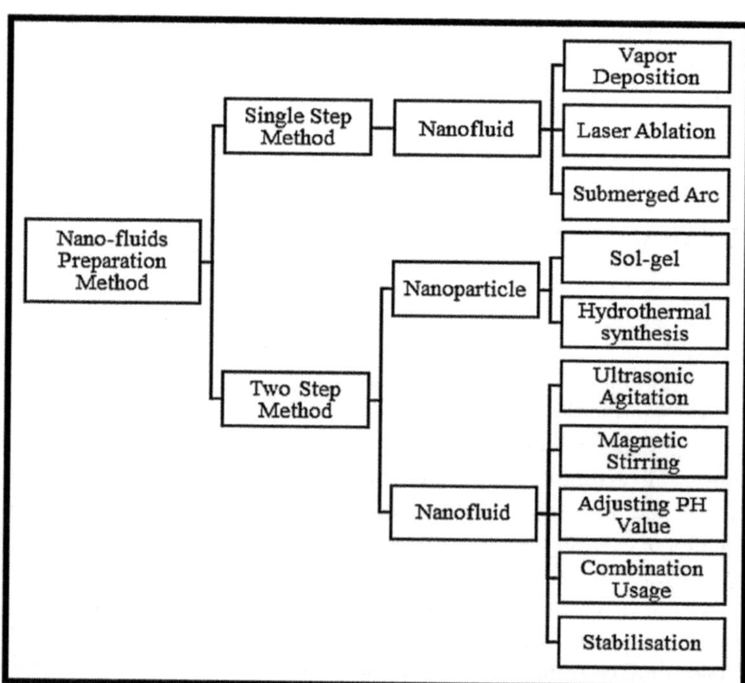

FIGURE 1.4 Preparation methodology of nanofluid.[8]

$$\text{Nanofluid concentration } (\varnothing), \text{Vol\%} = \frac{\left(\frac{\text{Weight}}{\text{Density}}\right)_{\text{Nanoparticle}}}{\left(\frac{\text{Weight}}{\text{Density}}\right)_{\text{Nanoparticle}} + \left(\frac{\text{Weight}}{\text{Density}}\right)_{\text{Base fluid}}} \times 100 \quad (1.1)$$

1.4 PRODUCTION OF HYBRID NANOFLUID

The production of hybrid nanofluids involves the combination of nanoparticles with different sizes and shapes inside a base fluid. The modification of essential features of hybrid nanofluids can be achieved by the adjustment of nanoparticle type and concentrations of individual elements. Consequently, these changes have a direct impact on the thermophysical characteristics of the fluid, including density, thermal conductivity, electrical conductivity, viscosity, and optical properties.

1.4.1 Preparation Methodology

The nanoparticles are incorporated into a base fluid, either separately or in a composite state, and then mixed together using magnetic stirring and high-intensity acoustic cavitation (**Figure 1.5**). The selection of nanoparticle type and base fluid is contingent upon the type of application. For example, the combination of water and ethylene glycol or water and propylene glycol can be blended to produce an antifreeze solution that is appropriate for colder environmental circumstances. Several studies have been undertaken to investigate the incorporation of single or hybrid nanoparticles into the base fluid. These investigations have confirmed that the resulting nanofluid demonstrates enhanced properties when compared to both the mono nanofluid and the base fluid.

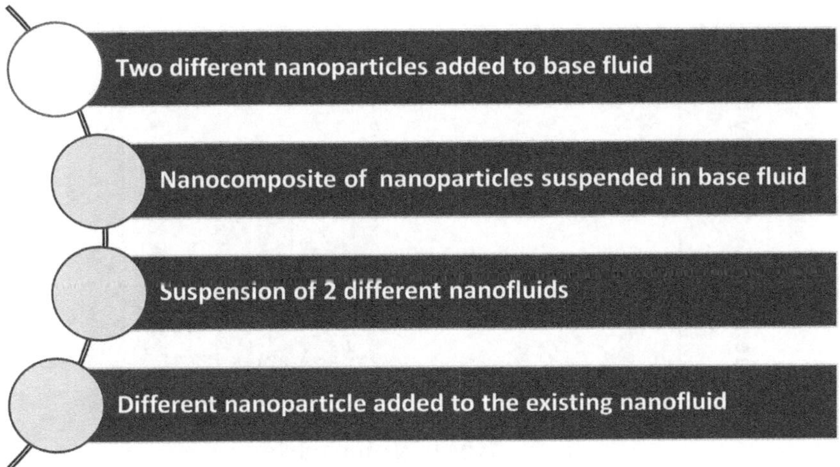

FIGURE 1.5 Preparation methodology of hybrid nanofluids.

Following the synthesis of hybrid nanoparticles, they undergo characterization procedures to assess their suitability for the intended uses. Subsequently, the nanoparticles are introduced into the base fluid in accordance with the desired solid volume fraction (φ, Vol%) from Equation (1.2). Here, *w*, *ρ*, and subscript *np1*, *np2*, and *bf* represent weight, density, nanoparticle 1, nanoparticle 2, and base fluid, respectively.

$$\varphi = \frac{\left(\frac{w}{\rho}\right)_{\text{np1}} + \left(\frac{w}{\rho}\right)_{\text{np2}}}{\left(\frac{w}{\rho}\right)_{\text{np1}} + \left(\frac{w}{\rho}\right)_{\text{np2}} + \left(\frac{w}{\rho}\right)_{\text{bf}}} \tag{1.2}$$

1.5 SIGNIFICANT FACTORS INFLUENCING THE CHARACTERISTICS OF HYBRID NANOFLUIDS

Several significant elements contribute to the properties demonstrated by nanofluids. After determining the specific type of nanoparticle to be incorporated, the researchers focused their efforts on determining the optimal concentration at which it can be introduced to the base fluid.

1.5.1 Type of Nanoparticle

Based on the application, choosing the appropriate nanoparticle is very important. The size, morphology, and thermophysical properties of the nanoparticles are some of the attributes that are included in this category. **Figure 1.6** provides an overview of a few of the nanoparticles that have garnered a lot of attention recently. These properties are able to be utilized within the context of their corresponding domains.

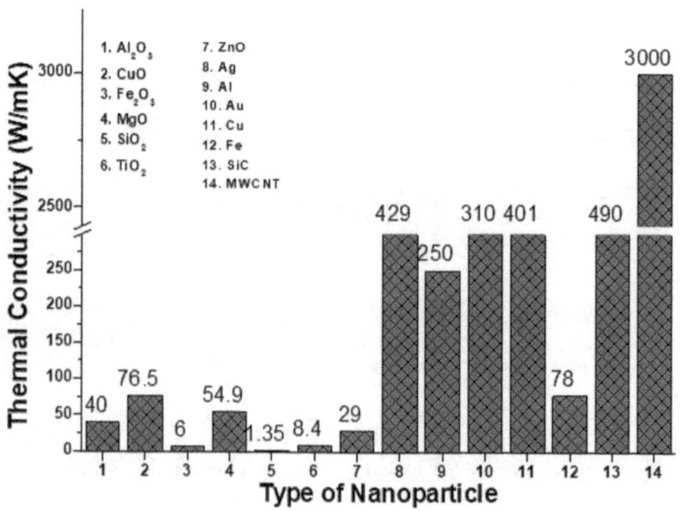

FIGURE 1.6 Thermal conductivity values of some of the nanoparticles.

The relevance of electrical conductivity as an element in the development of nanoscience is one of the most important aspects. When studying the stability of nanofluids, it is also important to take into consideration the morphology of the nanoparticles involved. On the contrary, the specific heat of nanoparticles is the factor that is responsible for various important characteristics. The quality that was just discussed is of critical significance in terms of the promotion and protection of thermal energy in the context of transportation and upkeep. The importance of nanoparticle density extends to a wide variety of domains, including pumping. The following table presents a variety of nanoparticles along with the characteristics that are unique to each one. The majority of academic research has supported the hypothesis that an increase in the thermal conductivity of nanoparticles can lead to an increase in the thermal conductivity of nanofluids. This overall logic is one of the primary drivers behind the utilization of nanofluids.

1.5.2 Type of Nanoparticle Combination

The preparation of a nanofluid necessitates careful consideration while combining nanoparticles. Nanofluid stability is regarded as a crucial parameter. A hybrid nanofluid is produced by the process of combining several nanoparticles and achieving homogeneity within a fluid medium. The nanoparticles exhibit a lack of interaction among themselves. The prevention of accumulation holds more significance. The matter of stability in hybrid nanofluids holds significant importance and warrants careful consideration. The selection of homogenous nanoparticles is crucial, and the hydrophilic or hydrophobic nature of these particles plays a significant role in this process. **Figure 1.7** illustrates the diverse range of hybrid nanofluids that have been

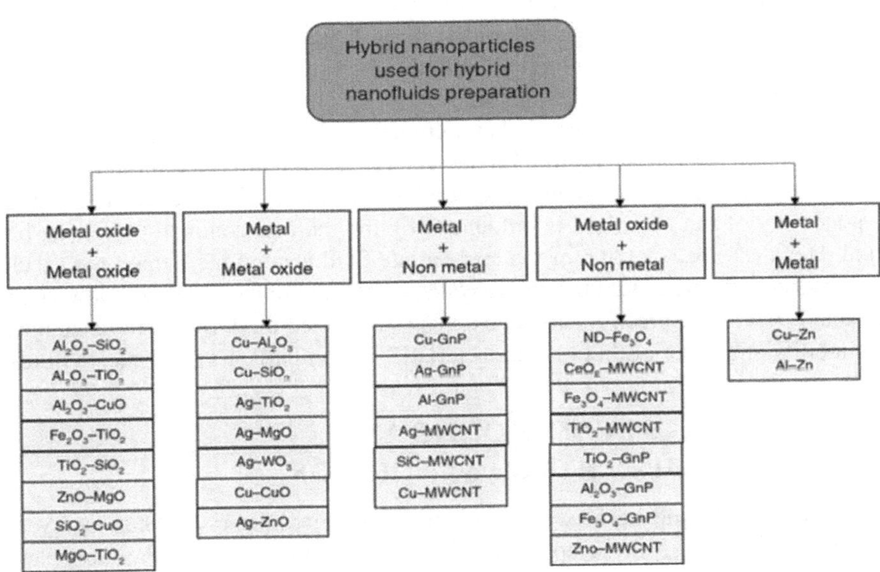

FIGURE 1.7 Various investigated hybrid nanofluid combinations.[9]

investigated by researchers to explore their potential applications. For example, carbon nanotubes are one of the most commonly employed nanoparticles in the synthesis of hybrid nanofluids due to their significant thermophysical properties.

1.5.3 The Concentration of Nanofluid

The majority of investigations have observed a linear rise in effective thermal conductivity as the concentration of nanofluids increases. In a stable suspension, the nanoparticles exhibit complete Brownian motion. The presence of a higher number of particles leads to an increased occurrence of thermal bridging phenomena, hence enhancing the relative characteristics of the resulting fluid. However, contrary to the findings, other writers have posited the possibility of particle agglomeration occurring at elevated concentrations of nanoparticles. Additionally, the quantity of nanoparticles added directly impacts the overall cost of the operation, necessitating the optimization of the concentration value of the nanofluid.

1.5.4 Type of Base Fluid

The selection of a suitable fluid is of utmost significance in the production of nanofluids. The properties of fluids play a significant role in determining the properties of nanofluids, thus necessitating the careful selection of an appropriate fluid based on its intended application. The viscosity of nanofluids exhibits a strong dependence on the type of fluid used. For example, commonly used base fluids like ethylene glycol have a viscosity of 18.376 cP, which is way higher than the water viscosity (0.890 cP). As the viscosity of the fluid increases, it becomes necessary to make adjustments to the pump performance to compensate for the increased resistance to shear. So, it is seen that there is a minor decrease in flow rate, a more notable decrease in head or pressure, and a major rise in power consumption. Consequently, the magnitude of the pumping power, which is contingent upon the level of shear stress, exhibits a strong correlation with the viscosity of the fluid.[10]

The type of fluid employed is also a crucial factor influencing the thermal conductivity of the nanofluid. **Figure 1.8** depicts a range of conventional fluids and their corresponding thermal conductivities, serving as a point of reference. The thermal conductivity of the nanofluid is influenced by the thermal conductivity of the base fluid. However, the selection of the appropriate fluid is contingent upon the specific application of the nanofluid inside the system.[11] Undoubtedly, water with low viscosity stands out as the preeminent and extensively utilized fluid, owing to its abundant availability and superior thermal characteristics in comparison to alternative fluids.[12]

1.6 CHALLENGES IN DEVELOPING AND IMPLEMENTING HYBRID NANOFLUIDS

At present, nanofluids are extensively employed in diverse applications; yet, a significant number of these applications remain confined to the laboratory setup. Nevertheless, the usage of nanofluids and hybrid nanofluids is constrained by several opportunities and constraints. The stability of hybrid nanofluids, which are

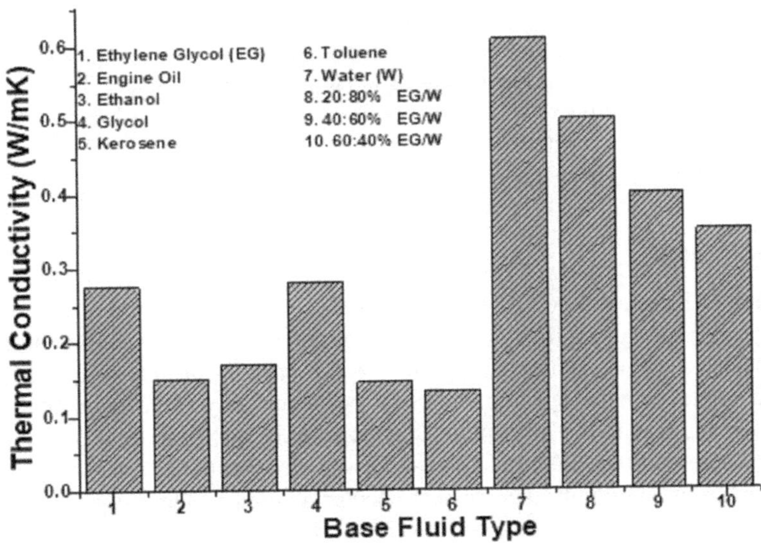

FIGURE 1.8 Thermal conductivity of conventional base fluids.

composed of several nanoparticles, is a significant barrier that hinders their widespread adoption in industrial applications. The incorporation of nanoparticles into hybrid nanofluids is a crucial factor in defining their stability, which is a fundamental aspect contributing to the overall benefits and improvement of heat transmission.[13] The presence of nanoparticles facilitates the development of aggregates due to the influence of robust van der Waals forces. Furthermore, a considerable number of researchers are currently engaged in the investigation of surface property modifications of nanoparticles using various methods such as functionalization, surfactant addition, or pH adjustment. The preservation of nanofluid characteristics despite the introduction of external agents remains crucial.

Nanofluids present significant theoretical challenges because of the observed experimental thermophysical properties that exceed the predictions made by current models. Currently, there is a lack of a comprehensive theoretical framework or equation that accounts for the thermophysical characteristics of nanofluids, specifically taking into consideration the distribution structures of uniformly dispersed nanoparticles within the base fluids. These distribution structures are known to have an impact on the thermal properties of nanofluids. Numerous research has been conducted to investigate the impact of various factors, such as nanoparticle size, shape, concentration, base fluid type, surfactant type and concentration, as well as temperature, on the performance of nanofluids. However, significant variations in outcomes can be observed even when employing identical nanofluids.[14] The proposed model equations are of significant importance in this context. However, the lack of model equations for hybrid nanofluids presents a challenge in accurately forecasting their properties. The industrialization of nanofluids can be facilitated by the provision of equations for the transportation and flow behavior of nanofluids.

Despite the shown efficacy of incorporating nanofluids, even at lower concentrations, in enhancing equipment performance, a contentious issue remains about the potential financial implications associated with this approach. To achieve this objective, it is imperative to set a price point for nanofluids that falls below a specific threshold. The price is intricately linked to the level of demand, which in turn is contingent upon the successful resolution of technological barriers.

1.7 SUMMARY

The field of nanotechnology has experienced significant growth in recent years, emerging as a multidisciplinary domain. A comprehensive understanding of the thermal, physical, chemical, optical, and magnetic properties of nanostructures has the potential to facilitate the development of various materials with practical applications. In light of this matter, the integration of nanoparticles into base fluids to address the issue of decreased thermal conductivity shown in conventional fluids has led to the development of nanofluids. The present chapter focuses on the origins and essential principles of hybrid nanofluids, in light of the recent achievements in this field of study. In conclusion, this study emphasizes the several aspects that impact the properties of hybrid nanofluids. These factors encompass the type, concentration, and combination of nanoparticles, as well as the type of the base fluid. The issues pertaining to the selection, combination, and mixing ratio of nanoparticles to produce hybrid nanofluids, as well as the absence of model equations to accurately predict the thermophysical properties of these nanofluids, have been emphasized.

REFERENCES

1. Nasrollahzadeh, M., Sajadi, S. M., Sajjadi, M., & Issaabadi, Z. (2019). An introduction to nanotechnology. *Interface Science and Technology*, *28*, 1–27. https://doi.org/10.1016/B978-0-12-813586-0.00001-8
2. Yadav, J., Jasrotia, P., Kashyap, P. L., Bhardwaj, A. K., Kumar, S., Singh, M., & Singh, G. P. (2022). Nanopesticides: Current status and scope for their application in agriculture. *Plant Protection Science*, *58*(1), 1–17. https://doi.org/10.17221/102/2020-PPS
3. Choi, S. (1998). Nanofluid technology: current status and future research. *Energy*, *26*. www.ncbi.nlm.nih.gov/pubmed/11048%5Cnwww.osti.gov/energycitations/product.biblio.jsp?osti_id=11048
4. Eastman, J. A., Choi, S. U. S., Li, S., Yu, W., & Thompson, L. J. (2001). Anomalously increased effective thermal conductivities of ethylene glycol-based nanofluids containing copper nanoparticles. *Applied Physics Letters*, *78*(6), 718–720. https://doi.org/10.1063/1.1341218
5. Keblinski, P., Phillpot, S. R., Choi, S. U. S., & Eastman, J. A. (2002). Mechanisms of heat flowin suspensions of nano-sized particles (nanofluids). *International Journal of Heat and Mass Transfer*, *45*, 855–863.
6. Olmo, C., Mendez, C., Ortiz, F., Delgado, F., Valiente, R., & Werle, P. (2019). Maghemite nanofluid based on natural ester: Cooling and insulation properties assessment. *IEEE Access*, *7*, 145851–145860. https://doi.org/10.1109/ACCESS.2019.2945547
7. Munkhbayar, B., Tanshen, M. R., Jeoun, J., Chung, H., & Jeong, H. (2013). Surfactant-free dispersion of silver nanoparticles into MWCNT-aqueous nanofluids prepared by one-step technique and their thermal characteristics. *Ceramics International*, *39*(6), 6415–6425. https://doi.org/10.1016/j.ceramint.2013.01.069

8. Ali, H. M., Babar, H., Shah, T. R., Sajid, M. U., Qasim, M. A., & Javed, S. (2018). Preparation techniques of TiO2 nanofluids and challenges: A review. *Applied Sciences, 8*(4), 1–30. https://doi.org/10.3390/app8040587
9. Alfellag, M. A., Kamar, H. M., Azwadi, N., Sidik, C., Ali, S., Kazi, S. N., Alawi, O. A., & Abidin, U. (2023). Rheological and thermophysical properties of hybrid nanofluids and their application in flat-plate solar collectors: A comprehensive review. *Journal of Thermal Analysis and Calorimetry, 148*, 6645–6686. https://doi.org/10.1007/s10973-023-12184-3
10. Malika, M., Pargaonkar, A., & Sonawane, S. S. (2024). Performance of an emulsion nanofluid membrane for the extraction of antimony Heavy Metal: Experimental and numerical investigation. *Journal of Sustainable Metallurgy*, 1–12.
11. Malika, M., Ashokkumar, M., & Sonawane, S. S. (2022). Mathematical and numerical investigations of nanofluid applications in the industrial heat exchangers. In *Applications of Nanofluids in Chemical and Bio-medical Process Industry* (pp. 53–78). Elsevier.
12. Malika, M., Thakur, P. P., & Sonawane, S. S. (2023). Sulfate/sulfur recovery from municipal wastewater treatment plants. In *Resource Recovery in Municipal Waste Waters* (pp. 145–164). Elsevier.
13. Thakur, P. P., Malika, M., & Sonawane, S. S. (2023). Energy recovery from industrial wastewaters. In *Resource Recovery in Industrial Waste Waters* (pp. 319–336). Elsevier.
14. Malika, M., & Sonawane, S. S. (2023). Few more significant applications of nanofluids. In *Nanofluid Applications for Advanced Thermal Solutions* (pp. 267–286). Elsevier.

2 Stability of Hybrid Nanofluids

2.1 INTRODUCTION

The attainment of enhanced properties in hybrid nanofluids relies on the physical and chemical compatibility of several nanoparticles within the base fluid, hence necessitating careful consideration. Two-step methodology is the common technique utilized for preparing hybrid nanofluids, where individual nanoparticles are added to the base fluid and homogenized using magnetic stirring and ultrasonic cavitation at high intensity. In most cases, if the nanoparticles are compatible together, there will be a stable solution formation. But, due to the versatile properties of nanoparticles, there is a need to adapt surface modification techniques including surfactant addition and pH modification or functionalization of nanoparticles. For example, due to their peculiar mechanical and physical properties (**Table 2.1**), multi-walled carbon nanotubes (MWCNTs) have been extensively investigated for their potential use in a wide range of industrial applications, such as photovoltaic thermal (PVT) energy storage systems, supercapacitors, sensors, and coolants. Furthermore, the significantly elevated thermal conductivity value of 3000 W/mK in comparison to water's thermal conductivity of 0.6 W/mK (**Figure 2.1**) has been observed.

TABLE 2.1
Properties of Multi-Walled Carbon Nanotube (MWCNT) Nanoparticles

Property	Value
Production method	Chemical vapor deposition (CVD)
Color	Black
Purity	> 90%
Average outer diameter	10–20 nm
Average inner diameter	5–10 nm
Average length	10–25 μm
True density	1000–2100 kg/m^3
Solubility in water	Not applicable
Electrical conductivity	100 Siemens per meter (S/m)
Thermal conductivity	1500–3000 W/mK
Specific surface area (SSA)	≈ 50–600 m^2/g
Content of COOH	≈ 2 Wt%
Ash content	< 1.5 Wt%
Physical property	Hydrophobic

DOI: 10.1201/9781003595137-2

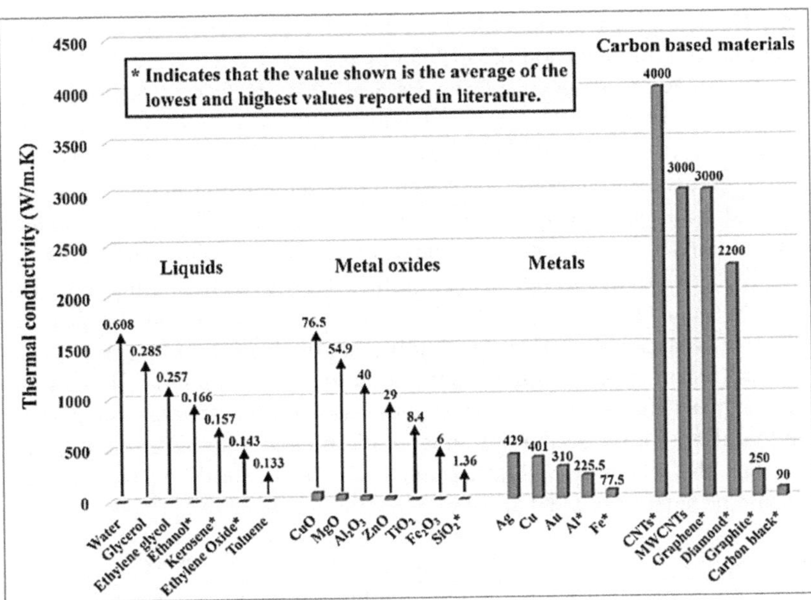

FIGURE 2.1 Thermal conductivity of multi-walled carbon nanotube (MWCNT) comparison over other commonly used base fluids and nanoparticles.[1]

Source: Ali N, Bahman AM, Aljuwayhel NF, Ebrahim SA, Mukherjee S, Alsayegh A. 2021. Carbon-based nanofluids and their advances toward heat transfer applications—a review. *Nanomaterials* [Internet]. 11(6):1628. doi:10.3390/nano11061628, open access.

In a study conducted by Sangeetha et al.[2] (2019), an experimental examination was carried out to evaluate the effects of nanofluids with varying concentrations of nanoparticles (namely, Al_2O_3, CuO, and MWCNT) in water on the efficiency of PVT systems. In general, solar power can be converted into usable electricity via photovoltaic, also known as PV, systems, exhibiting an efficiency of less than 20%. In this context, the rise in cell temperature resulting from the absorption of heat led to a reduction in the electrical conversion efficiency. To mitigate the issue of PV module heating, scholars have implemented a PV/T system, which incorporates a coolant to diminish the amount of heat created. The three distinct nanofluids were employed at varying concentrations (0%, 0.5%, 1%, 1.5%, and 2.0% by volume) in the capacity of coolants. The trials were primarily conducted throughout the hours of 9:30 am to 4:30 pm. The study's findings revealed that the application of commercially manufactured nano-emulsions led to improved performance of the PVT system. This was demonstrated by a notable improvement in system efficiency compared to the utilization of water only for cooling. The study's findings suggest that MWCNTs provide enhanced cooling capabilities in PV/T systems when compared to copper oxide (CuO) and aluminum oxide (Al_2O_3) particles.

In general, the hydrophobic character of MWCNTs can be attributed to the high intermolecular van der Waals interactions between the carbon atoms. Consequently, achieving dispersion of MWCNTs in water at ambient temperature is typically not feasible. Numerous researchers have conducted experiments on surface modification techniques. The inclusion of surfactants has been extensively investigated as a method to achieve a stable suspension of MWCNTs in water-based nanofluids. In this context, Pourpasha et al.[3] included Triton X-100 surfactant into the turbine oil base fluid at a ratio of 1:3. To ensure long-term stability of nano-lubricants, researchers have employed the techniques of homogenization using magnetic stirring and high-intensity ultrasonication. In a separate investigation, Choi et al.[4] conducted an experiment to examine the impact of incorporating four distinct surfactants, namely Sodium Dodecyl Benzene Sulfonate (SDBS), Cetyltrimethylammonium Bromide (CTAB), Sodium Dodecyl Sulfate (SDS), and Triton X-100, on the dispersion of MWCNTs in water. The suspension stability factor (**Equation 2.1**) was determined using laser transmission equipment after a period of one month following the production of the surfactant-assisted nanofluid. Based on the findings, the authors concluded that the stability period of the nanofluid treated with SDBS and TX-100 was greater than that of the nanofluid treated with CTAB and SDS, without explicitly discussing the surfactant mechanism.

$$\begin{matrix}\text{Suspension}\\ \text{stability factor}\end{matrix} = \frac{(\text{Light transmitted intensity})_{\text{Cuvette}} - (\text{Light transmitted intensity})_{\text{Nanofluid, Day 30}}}{(\text{Light transmitted intensity})_{\text{Cuvette}} - (\text{Light transmitted intensity})_{\text{Nanofluid, Day 0}}} \quad (2.1)$$

Multiple research papers have provided evidence supporting the theory that the agglomeration of particles can be influenced by the van der Waals and cohesive forces acting between particles. The aggregation of nanoparticles results in the sedimentation and blockage of pipes through which they are moving, as well as a decrease in the thermophysical characteristics of nanofluids. The assessment of stability is a fundamental factor that influences the properties of nanofluids in real-world applications. Hence, it is crucial to conduct a thorough investigation and analysis of the variables that impact the stability of nanofluid dispersion. Chapter 1 of the study delves into the process of preparing nanofluids and hybrid nanofluids. Subsequent chapters will focus on elucidating the mechanisms employed to enhance the stability of hybrid nanofluids and the methodologies utilized for their evaluation.

2.2 STABILITY ENHANCEMENT METHODOLOGIES

The absence of sufficient stability in hybrid nanofluids might lead to alterations in the thermophysical properties, thus impacting the rate of enhancement. So, the expected properties from the hybrid nanofluids can be achieved only when the particles are completely in the Brownian motion. When the nanoparticle is dispersed into the base fluid, many microscopic forces will be acting on it. When the particles in the Brownian motion will counterbalance the gravitational force, the particles will disperse in the base fluid for a longer period. The frequency of particle interactions is contingent upon the concentration of the nanofluid, whereby higher concentrations result in shorter interparticle distances. When the particle attraction force outweighs the repelling force within a specified distance, nanoparticles tend to cluster. Additionally, the morphology of nanoparticles plays a crucial role in comprehending the interparticle proximity. For instance, nanoparticles with rod, plate, or tube shapes exhibit a greater contact area compared to other shapes. Consequently, there exists a potential for enhanced attraction interactions between the particles, leading to the formation of larger aggregates. Instead of undergoing complete dispersion, particles will agglomerate and settle. In the context of thermal applications, it is observed that, when the temperature of the nanofluid rises, the nanoparticles inside the fluid absorb energy, leading to an acceleration of their Brownian motion. Moreover, the viscosity of the fluid will decrease, potentially leading to the agglomeration of particles.

2.2.1 Mechanical Nanofluid Stabilization Techniques

There are multiple techniques that can be employed to improve the stability of nanofluids, such as mechanical or chemical approaches, as depicted in **Figure 2.2**. Mechanical methods employed for stabilizing nanofluids involve the utilization of several techniques such as magnetic stirring, bath or probe ultrasonication, and high-pressure homogenization. These methods facilitate the homogenization of nanoparticles within the base fluid. The utilization of magnetic stirring is a highly adaptable method for achieving homogeneity within samples, regardless of whether they consist of identical or distinct phases. In the two-step approach for nanofluid preparation, the magnetic stirring technique is employed to facilitate the homogenous dispersion of nanoparticles within the base fluid. According to **Figure 2.3**, a typical magnetic stirrer is equipped with a pair of knobs. The right knob is utilized to adjust the rotational speed (measured in RPM), while the magnetic field remains constant. The rotation speed of the magnetic bar can be incrementally raised in response to the volume of fluid contained within the beaker. Alternatively, the temperature of the magnetic stirrer's left knob can be regulated to enhance the mixing of

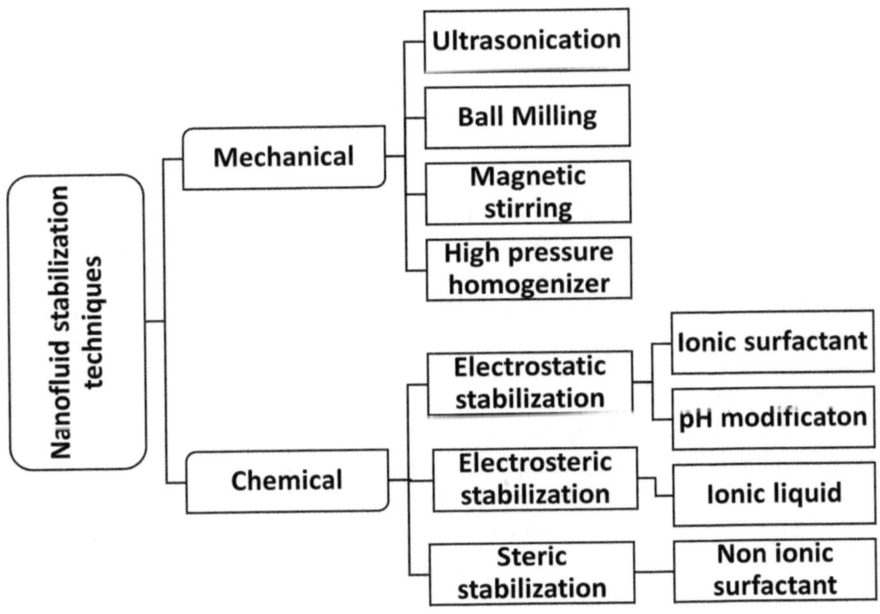

FIGURE 2.2 Presently employed methods for stabilizing nanofluids.

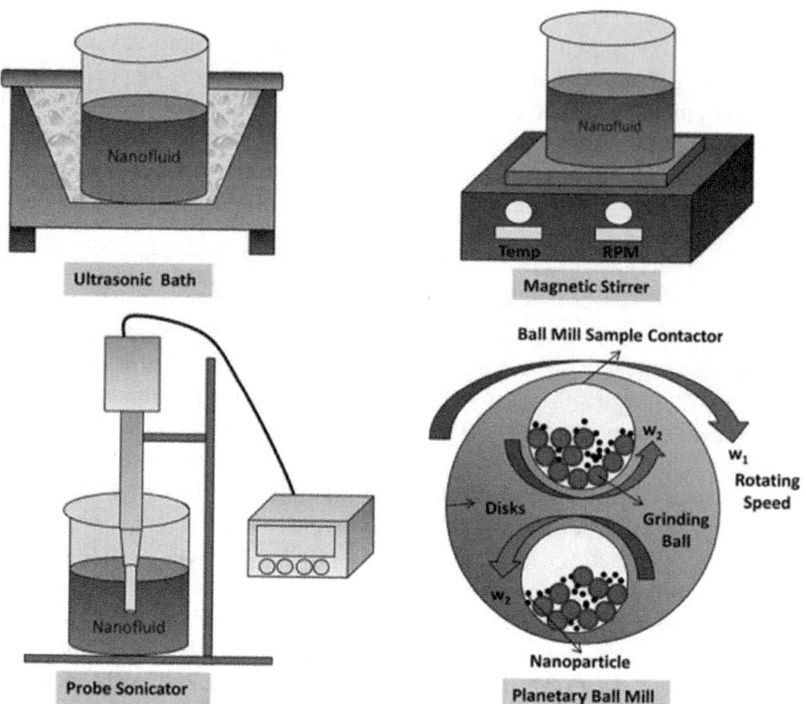

FIGURE 2.3 Stabilization of nanofluids via mechanical means.[5]

Source: Chakraborty S, Panigrahi PK. 2020. Stability of nanofluid: A review. *Appl Therm Eng*. 174:115259. doi:10.1016/J.APPLTHERMALENG.2020.115259

nanoparticles. To enhance the stability duration, the process of magnetic stirring is then employed in conjunction with ultrasonication.

Ultrasonication is a commonly utilized physical technique for the dispersion of nanoparticles in base fluids. The sonication approach is found to yield superior dispersion, exhibiting a reduced likelihood of particle agglomeration compared to magnetic stirring. The sonication method utilizes ultrasonic waves to provide energy to a nanofluid. The utilization of sonication facilitates the development of a more homogenous suspension. In terms of achieving consistent dispersion, sonication is a more effective method compared to the use of a magnetic stirrer. Ultrasound waves can be propagated through the utilization of either a bath-type or probe-type sonicator. In the experimental setup seen in **Figure 2.3**, the sonicator is immersed in a water bath. The ultrasonic waves are initially transmitted through the water bath and subsequently directed toward the glass container containing the nanofluid. The bath sonicator exhibits minimal sonic cavitation due to the indirect transmission of ultrasonic energy into the fluid. On the contrary, when the tip of the probe is put directly into the fluid medium, the probe sonicator offers more control and precision. The tip can be directed toward certain tasks such as homogenization, dispersion, and deagglomeration of particles.

The liquid medium experiences the transmission of sound waves by the introduction of high-power ultrasound. This results in the formation of alternating cycles of high and low pressure inside the fluid determining the rate of these cycles. According to the cavitation principle, vacuum bubbles are generated in the liquid at low-pressure conditions. Due to sudden expansion and contraction, bubbles undergo a rapid collapse during a cycle of high pressure. The occurrence in question is commonly referred to as cavitation. During the process of implosion, localized conditions of extremely high temperatures (about 5,000 Kelvin) and pressures (approximately 2,000 atmospheres) are attained.

High pressure can serve as a means to disperse the nanoparticles and hence to mitigate their agglomeration. The phenomenon of freed temperature has the potential to induce localized heating. Factors such as local heating, surface energy, van der Waals forces, and electrostatic interactions could further contribute to the clustering of particles. To mitigate this issue, the probe sonicators are operated utilizing a pulse mode. The probe sonicator has the capability to be operated in two distinct modes: pulse mode and continuous mode. In the context of continuous mode operation involving the transmission of continuous sonic waves, there exists a potential for localized thermal elevation. However, in the pulse mode, sonic waves are transmitted at intervals of either 10 or 20 seconds. This allows sufficient time for the temperature to evenly distribute throughout the fluid.

The utilization of high-shear and high-pressure homogenizers is widely acknowledged as an efficient approach to enhance the stability of nanofluids by disintegrating clusters into their constituent particle sizes. In certain instances, it is common practice to subject nanoparticles to ball milling before their dispersion in the base fluid. This milling process serves to significantly decrease the average cluster size of the particles and hence improves the stability of the resulting nanofluid. Nevertheless, this methodology is unable to sustain the stability of nanofluids for extended periods and often proves to be useless when dealing with larger particle concentrations.

2.2.2 Chemical Nanofluid Stabilization Techniques

The selection of an appropriate type of nanoparticle and base fluid is crucial as it significantly influences the overall properties of the resulting hybrid nanofluid. In general, fluids can be classified into two categories: polar and non-polar, while nanoparticles can be categorized as either hydrophilic or hydrophobic. Hydrophilic nanoparticles can readily disperse in aqueous solutions using straightforward homogenization methods such as magnetic stirring and ultrasonication. Aluminum oxide (Al_2O_3), titanium dioxide (TiO_2), copper oxide (CuO), silica (silicon dioxide, SiO_2), zinc oxide (ZnO), gold (Au), graphene oxide nanosheets, and magnesium oxide (MgO) nanoparticles exhibit hydrophilic properties. Silver (Ag), carbon nanotubes (CNT—multi-walled (MWCNT), single-walled (SWCNT)), siloxane groups, graphene, graphite, carbon black, iron (III) oxide (Fe_2O_3), iron (II) oxide (FeO), and magnetite (Fe_3O_4) nanoparticles are hydrophobic. When dealing with hydrophobic nanoparticles, it is more advantageous to disperse them in non-polar liquids without relying on external stabilizing agents. To achieve the synthesis of a stable nanofluid, it is crucial to utilize a variety of stabilizing approaches rooted in chemistry. These techniques entail the amalgamation of hydrophilic nanoparticles with non-polar solvents or, conversely, the mixing of hydrophobic nanoparticles with polar solvents. Additionally, physical stabilizing methods are employed in this process.

As seen in **Figure 2.2** there are three main categories of chemical stabilizing approaches for nanofluids: electrostatic stabilization, steric stabilization, and electrosteric stabilization.

2.2.2.1 Electrostatic Stabilization

Electrostatic stabilization of nanofluid can be accomplished by altering the overall surface charge of the nanofluid, which includes adsorption or substitution or detachment of ions, charge adsorption on the nanoparticle surface, segregation, or reduction of electrons (e^-) at the nanoparticle surface. This process operates based on the theory of electrostatic repulsion between two particles, which is a result of their surface charges. When two particles come into proximity under operational circumstances, there is a phenomenon known as double-layer overlapping, which results in the emergence of a repulsive force between the nanoparticles of similar size.

Nanoparticles can exhibit either anionic or cationic properties. Upon introduction into a solution, the nanoparticles interact with the ions present in the solution, leading to adsorption or rejection processes. These interactions are influenced by the pH of the solution, ultimately resulting in the acquisition of a positive or negative surface charge of the nanoparticle, as depicted in **Figure 2.4(a)**. In the presence of an electric field, oppositely charged ions are drawn toward the surface of the nanoparticle, resulting in the formation of an electrical double layer, as depicted in **Figure 2.4(b)**. The concentration of ions with opposite charges may decrease as the distance increases. When two charged surfaces come into proximity and their ion clouds start to overlap, a repulsive contact known as a double-layer interaction occurs (see **Figure 2.4(c)**). To summarize, in electrostatic stabilization, the stability of particles can be achieved using ionic substances, which involve pH adjustment and the utilization of cationic/anionic surfactants. On the contrary, non-ionic substances, such as polymers and non-ionic surfactants, are employed for electro-steric stabilization. Due to its hydrophobic nature, the MWCNT exhibits limited dispersibility in polar liquids in the absence of a dispersion mechanism. The observed superhydrophobicity, as depicted in **Figure 2.5**, results in the material exhibiting a tendency to float readily on the surface of the water. In a recent paper, Arya et al.[6] analyzed the performance of a nanofluid consisting of MWCNT dispersed in water, specifically in the context of a flat plate heat exchanger. In this

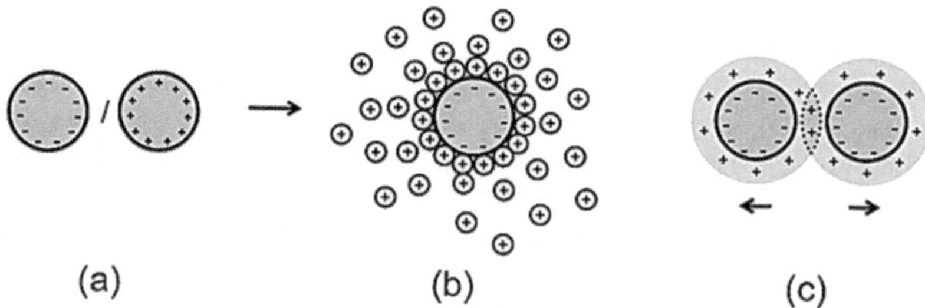

FIGURE 2.4 Electrostatic stabilization.[7] (a) Nanoparticle (positively or negatively charged) (b) Double-layer formation from the adsorption or substitution or detachment of ions, charge adsorption on the nanoparticle surface, segregation, or reduction of electrons (e^-) at the nanoparticle surface. (c) Stable dispersion due to electrostatic repulsion between the particles.

Source: Matter F, Luna AL, Niederberger M. 2020. From colloidal dispersions to aerogels: How to master nanoparticle gelation. *Nano Today*. 30:100827. doi:10.1016/J.NANTOD.2019.100827: open access

context, the author employed the electrostatic stabilization technique to enhance the dispersion of MWCNT nanoparticles in water. This was achieved by adjusting the pH of the solution, as depicted in **Figure 2.6**. To adjust the pH of the solution, solutions of NaOH and HCl with concentrations of 0.1 mM were employed. The electrostatic potential differential between a nanoparticle and the surrounding bulk solution is a metric used to quantify the disparity between their respective

FIGURE 2.5 Multi-walled carbon nanotube (MWCNT) nanoparticles (left) and MWCNT in water without any dispersion mechanism (right).

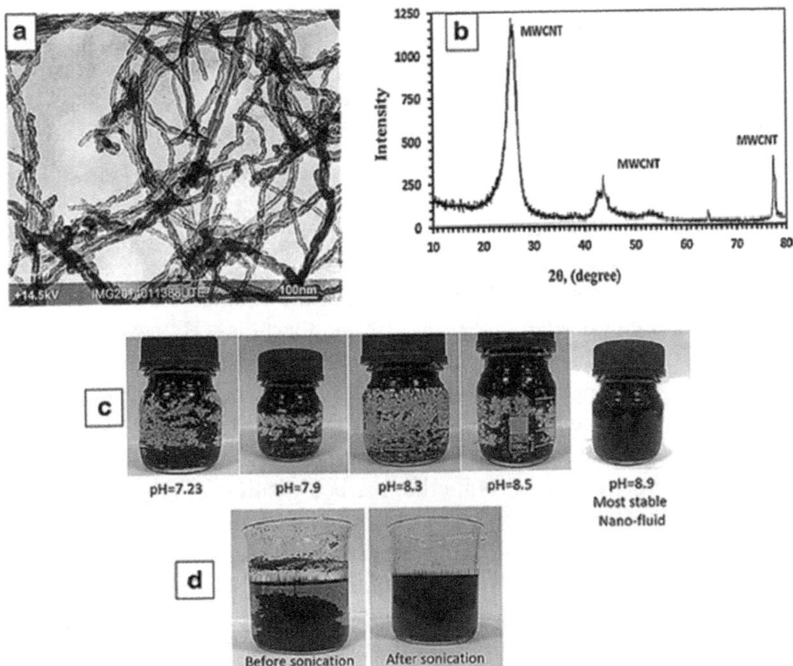

FIGURE 2.6 Characterization of 0.2 Wt% multi-walled carbon nanotube (MWCNT) in water nanofluid.[6] (a) Transmission electron microscopic (TEM) image. (b) X-ray fluorescence (XRF). (c) Effect of pH on the stability of nanofluid. (d) Effect of ultrasonication on the stability of nanofluid.

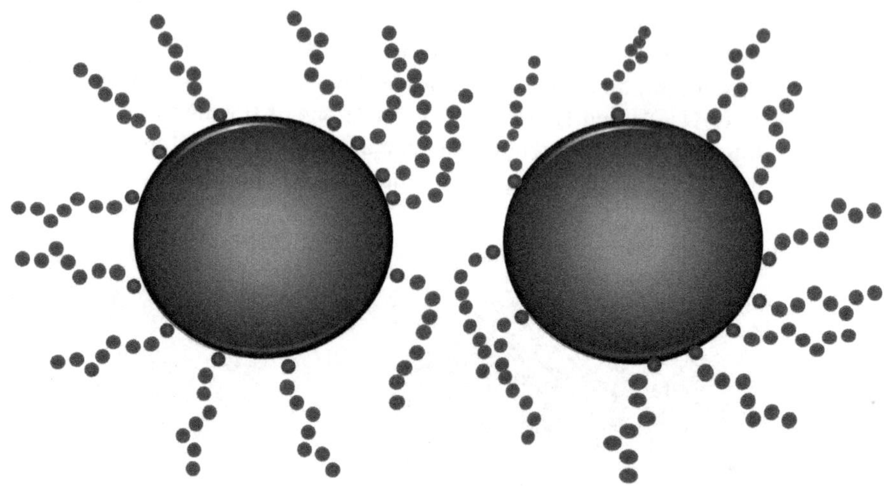

FIGURE 2.7 Electro-steric stabilization of nanofluids.[8]

Source: López-Esparza R, Altamirano MAB, Pérez E, Goicochea AG. 2015. Importance of molecular interactions in colloidal dispersions. *Adv Condens Matter Phys.* 2015:1–8. doi: http://dx.doi.org/10.1155/2015/683716 Review; open access.

potentials. The generation and quantification of electrostatic potential in terms of zetapotential (mV) are attributed to the surface energy of nanoparticles present in a base fluid. The zetapotential value was seen to vary between 30 and 45 mV when the pH was adjusted to 8.9, indicating an increased level of colloidal stability.

2.2.2.2 Steric Stabilization

Steric stabilization is achieved by attaching substantial molecules including non-ionic surfactants and long polymer chains to the nanoparticle surface. By the attachment of the external molecules, the prevention of the van der Waals attractive forces is reduced amid the colloidal suspension of nanoparticles. However, it is predicted that the polymeric chain and nanoparticle should have a strong affinity to create steric stabilization. The longer the polymeric layer chain, the bigger the space between the particles to protect the nanofluid from aggregation (**Figure 2.7**). As this phenomenon is not a function of ions, steric stabilization can also be used for nonaqueous solvents including benzene, alcohol, ether, carbon disulfide, and acetone. As shown in Figure 2.7, the head of the polymer is adsorbed onto the nanoparticle surface and the tail interacts with the solvent. Moreover, the polymer chain acts as a protective layer to avoid the agglomeration of the particles.

Either chemisorption or physisorption can be used to attach the monomer or co-polymer to the surface of the nanoparticle.[9] The chemisorption results in a stronger adherence of the polymer to the surface of the nanoparticle than would otherwise be the case. In addition, amphiphilic surfactants have the ability to generate steric stabilization in polar and non-polar solvents. This is a significant advantage. Several examples of the generation of steric stabilization are listed in **Table 2.2**.

TABLE 2.2
Examples of Non-Ionic Steric Stabilization

Polymer (Head–Tail) Combination	Non-Ionic Surfactant
Polystyrene $(C_8H_8)_n$–poly (oxy ethylene)/PEG	Gum Arabic (natural gum)
Polyvinyl chloride (PVC)–polyvinyl pyrrolidone (PVP)	Poly-vinylpyrrolidone
Poly (1-acetyloxiethylene)—poly (vinyl alcohol)	Triton X-100
Poly (propylene)-poly (vinyl methyl)	Tween 20 and Span 80

2.2.2.3 Electro-Steric Stabilization

Electro-steric stabilization refers to the concurrent utilization of electrostatic and steric stabilization mechanisms. Instead of employing a non-ionic polymer, the surface of the nanoparticle is coated with an ionic polymer.[10] The choice of protective polymer layer (steric interference) and the electric potential difference (electrostatic repulsion) between nanoparticles can be determined based on the nanoparticle's type (cationic/anionic) and the appropriate ionic polymer. The polymer's head is adsorbed onto the surface of the nanoparticle, while its tail interacts with the solvent.[11] This interaction gives rise to steric and electrostatic repulsions between the particles. Moreover, the polymer chain functions as a protective shield to hinder the aggregation of particles. The suitable electro-steric stabilizers are ionizable polymer chains containing carboxylic (–COOH) and sulfonate ($-SO_3H$) groups. The adsorption of polymer ends is facilitated by the electrostatic interaction between the charged head of the polymer and the surface of the nanoparticle.[12] Ionic liquids that consist of a combination of organic cations and either inorganic or organic anions have been found to be highly effective as electro-steric stabilizers (**Figure 2.8**).

Once the nanofluid is stabilized using electrostatic, steric, and electro-steric techniques, its stability can be evaluated in terms of zetapotential (ζ milli Volt/mV), absorbance fluctuation measured by a UV spectrometer, dynamic light scattering, centrifugation, and sedimentation analysis.

2.3 SUMMARY AND FUTURE REMARKS

The successful industrial implementation of nanofluids depends substantially on the pivotal feature of long-term stability. To attain the desired enhancement in thermophysical characteristics, it is important to conduct methodical and thorough laboratory and pilot size investigations aimed at comprehending the process factors. In addition, in the case of hybrid nanofluids, it is crucial to consider not only the compatibility between the nanoparticle and base fluid but also the maintenance of compatibility between the nanoparticle itself. For instance, hydrophilic nanoparticles exhibit a high degree of dispersibility in polar solvents without the need for external agents. However, achieving the dispersion of hydrophobic nanoparticles in polar solvents is a more complex task. In this scenario, it is necessary to employ various approaches for stabilization, including electrostatic stabilization, steric stabilization,

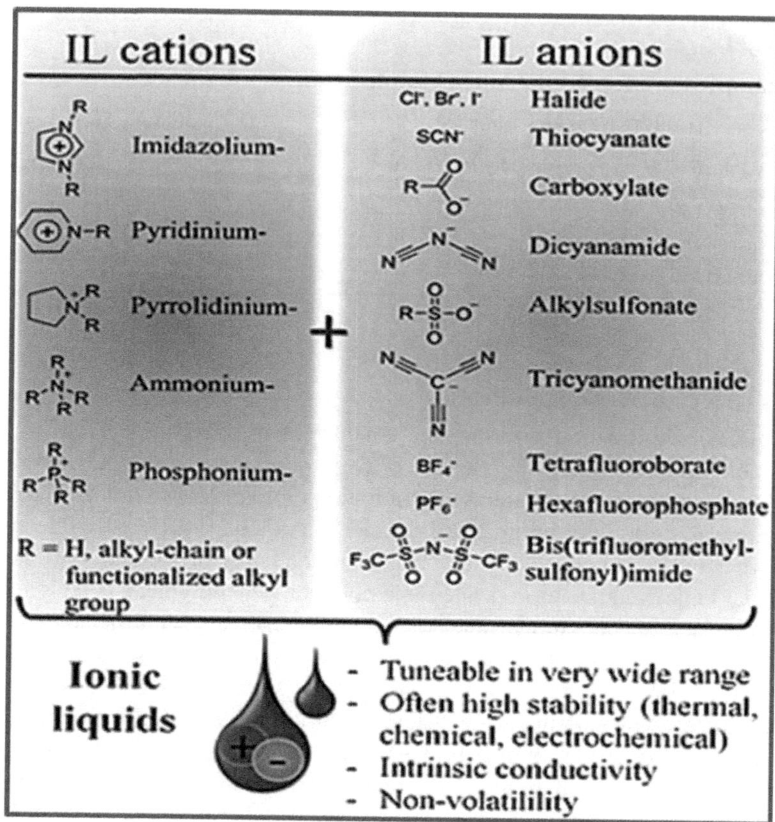

FIGURE 2.8 Different types of ionic liquids used for electro-steric stabilization.[13]

Source: Rauber, Daniel, Frederik Philippi, Johann Seibert, Johannes Huwer, Harald Natter, and Rolf Hempelmann. From current science to school—the facets of green chemistry on the example of ionic liquids. *World J Chem Edu.* 7, no. 2 (2019): 153–165; open access.

and electro-steric stabilization. However, the selection of a suitable methodology can be accomplished by either a trial-and-error approach or a comprehensive review of the existing literature. All three approaches employ either ionic or non-ionic chemicals, such as surfactants, polymer chains, or ionic liquids, for adsorption onto the surface of the nanoparticles. Furthermore, the current body of research on the application of these approaches for the stabilization of hybrid nanofluids is restricted. Additional research is necessary to gain a comprehensive understanding of the behavior of nanoparticles within the given experimental conditions.

REFERENCES

1. Ali, N., Bahman, A. M., Aljuwayhel, N. F., Ebrahim, S. A., Mukherjee, S., & Alsayegh, A. (2021). Carbon-based nanofluids and their advances towards heat transfer applications—a review. *Nanomaterials, 11*(6), 1628. https://doi.org/10.3390/nano11061628

2. Sangeetha, M., & Manigandan, S. (2019, May). Progress of MWCNT, Al2O3, and CuO with water in enhancing the photovoltaic thermal system. *International Journal of Energy Research*, 1–12. https://doi.org/10.1002/er.4905
3. Pourpasha, H., Zeinali Heris, S., Mahian, O., & Wongwises, S. (2020). The effect of multi-wall carbon nanotubes/turbine meter oil nanofluid concentration on the thermophysical properties of lubricants. *Powder Technology, 367*, 133–142. https://doi.org/10.1016/J.POWTEC.2020.03.037
4. Choi, T. J., Jang, S. P., & Kedzierski, M. A. (2018). Effect of surfactants on the stability and solar thermal absorption characteristics of water-based nanofluids with multi-walled carbon nanotubes. *International Journal of Heat and Mass Transfer, 122*, 483–490. https://doi.org/10.1016/j.ijheatmasstransfer.2018.01.141
5. Chakraborty, S., & Panigrahi, P. K. (2020). Stability of nanofluid: A review. *Applied Thermal Engineering, 174*, 115259. https://doi.org/10.1016/J.APPLTHERMALENG.2020.115259
6. Arya, A., Sarafraz, M. M., Shahmiri, S., Madani, S. A. H., Nikkhah, V., & Nakhjavani, S. M. (2018). Thermal performance analysis of a flat heat pipe working with carbon nanotube-water nanofluid for cooling of a high heat flux heater. *Heat and Mass Transfer, 54*, 985–997. https://doi.org/10.1007/s00231-017-2201-6
7. Matter, F., Luna, A. L., & Niederberger, M. (2020). From colloidal dispersions to aerogels: How to master nanoparticle gelation. *Nano Today, 30*, 100827. https://doi.org/10.1016/J.NANTOD.2019.100827
8. López-Esparza, R., Altamirano, M. A. B., Pérez, E., & Goicochea, A. G. (2015). Importance of molecular interactions in colloidal dispersions. *Advances in Condensed Matter Physics, 2015*, 1–8. http://doi.org/10.1155/2015/683716
9. Krishnan, S. S. J., Malika, M., Sharifpur, M., Sonawane, S. S., Mahian, O., & Meyer, J. P. (2023). Progress and challenges in nanofluids research. In *Nanofluid Applications for Advanced Thermal Solutions* (pp. 327–348). Elsevier.
10. Malika, M., & Sonawane, S. S. (2022). Experimental investigation of nanofluid in industrial heat exchangers. In *Applications of Nanofluids in Chemical and Bio-medical Process Industry* (pp. 79–106). Elsevier.
11. Malika, M. M., & Sonawane, S. S. (2019). Review on application of nanofluid/nano particle as water disinfectant. *Journal of Indian Association for Environmental Management (JIAEM), 39*(1–4), 21–24.
12. Malika, M., & Sonawane, S. S. (2020). Effect of nanoparticle mixed ratio on stability and thermo-physical properties of CuO-ZnO/water-based hybrid nanofluid. *Journal of the Indian Chemical Society, 97*(3), 414–419.
13. Rauber, D., Philippi, F., Seibert, J., Huwer, J., Natter, H., & Hempelmann, R. (2019). From current science to school—the facets of green chemistry on the example of Ionic liquids. *World Journal of Chemical Education, 7*(2), 153–165. https://doi.org/10.12691/wjce-7-2-15

3 Thermophysical, Optical, Electrical, and Magnetic Properties of Hybrid Nanofluids

Upon dispersing nanoparticles in the base fluid, the fluid undergoes a noticeable shift in behavior, resulting in alterations to its thermophysical, optical, electrical, magnetic, and dielectric properties. The degree of enhancement mostly depends on the nanoparticle type, base fluid, and the concentration of suspended nanoparticles in the nanofluid. Furthermore, when it comes to hybrid nanofluids (HNFs), the compatibility of nanoparticles and the ratio at which they are mixed are crucial considerations. In addition, researchers are conducting experimental measurements to determine the relevant qualities of the end application. In circumstances where experimentation is not feasible, researchers are resorting to extrapolation using existing theoretical models. When the HNF is used as a coolant, its thermophysical parameters such as thermal conductivity, viscosity, density, heat capacity, and surface tension are determined. To utilize nanofluid in solar collector applications, it is necessary to measure its optical properties, such as the extinction coefficient and its capacity to absorb incident light.

3.1 THERMOPHYSICAL PROPERTIES

3.1.1 Thermal Conductivity (k)

Thermal conductivity, represented by the symbol "k", is a measurement of the ability of a material to conduct or transport heat and is expressed in units of $\left[\frac{\text{Watt}}{\text{m.K}}\right]$, which is part of the International System of Units (SI). The material's heat conduction strength increases as the "k" value increases. Gases generally exhibit lower thermal conductivity compared to solids. In addition, pure metals with a significant number of free electrons have better thermal conductivity compared to other solids. As a result of the short distance between intermolecular gaps and their increased kinetic energy, they are densely arranged. Hence, the thermal energy transmission efficiency in liquids and gases is comparatively lower than that in solids. At room temperature, the thermal conductivity of multi-walled carbon nanotubes (MWCNT) is more than 3000 times higher than that of water (**Figure 3.1**).

The thermal conductivity of a working fluid in a thermal system is considered the most essential thermophysical attribute. Enhancing the thermal conductivity of

DOI: 10.1201/9781003595137-3

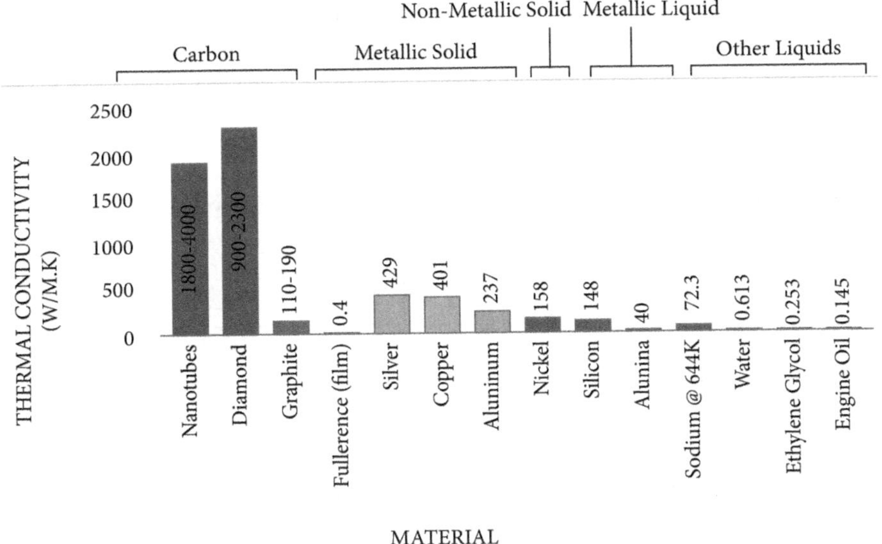

FIGURE 3.1 Thermal conductivity of various materials at 25°C.[1]

Source: Ukueje WE, Abam FI, Obi A. A perspective review on thermal conductivity of hybrid nanofluids and their application in automobile radiator cooling. *J Nanotechnol.* 2022. Open access.

a working fluid increases the total heat transfer rate, hence improving the overall efficiency of the thermal system. Thermal conductivity in materials science refers to the ability of a substance to transfer heat. A strong association exists between temperature and both the composition and thermal conductivity of the working fluid. Conventional thermal systems commonly employ working fluids obtained from typical base fluids, such as water and ethylene glycol (EG). Modern thermal systems currently employ working solutions composed of nanoparticles, thanks to the progress made in nanotechnology. Nanofluids, a distinct category of nanoparticles, are employed in certain practical solutions. HNFs, composed of nanoparticles derived from several nanomaterials, offer alternative and efficient solutions. Incorporating nanofluids, particularly HFs, into conventional base fluids has been observed to greatly improve the thermal conductivity of the working solution. The thermal conductivity of an HNF is determined by various factors, such as the mass fraction and nature of nanoparticles, the composition of the base fluid(s), and the volume concentration of the nanoparticles. It is increased as a result of the intensified Brownian motion occurring within the solution, which contains nanoparticles. During this event, the probability of chain creation and, consequently, the rate of interaction between the base fluid and the nanoparticles it contains are increased.

This is due to the simultaneous occurrence of enhanced Brownian motion and changes in kinetic energy, which are accompanied by an increased rate of molecular collisions. The frequency of lattice vibration increases due to the reduced particle separation resulting from enhanced contact between the base fluid and the nanoparticles. Due to the absence of theoretical model equations for forecasting the thermal conductivity of HNFs, the authors initially employed a modified Maxwell equation. However, upon doing direct experimental measurements, discrepancies were observed in the obtained results. To overcome these limitations, the authors have introduced a novel mathematical equation to describe the combination of nanoparticles that were tested in the experiment. Here, φ represents the concentration of nanofluid, while *k* represents thermal conductivity. The subscripts *Np1, bf,* and *Np2* denote nanoparticle 1, base fluid, and nanoparticle 2, respectively.

$$k_{HNF} = \frac{\frac{\left(\varnothing_{\mathrm{Np1}} k_{\mathrm{Np1}} + \varnothing_{\mathrm{Np1}} k_{\mathrm{Np1}}\right)}{\varnothing} + 2k_{\mathrm{bf}} + 2\left(\varnothing_{\mathrm{Np1}} k_{\mathrm{Np1}} + \varnothing_{\mathrm{Np2}} k_{\mathrm{Np2}}\right) - 2\varnothing k_{\mathrm{bf}}}{\frac{\left(\varnothing_{\mathrm{Np1}} k_{\mathrm{Np1}} + \varnothing_{\mathrm{Np1}} k_{\mathrm{Np1}}\right)}{\varnothing} + 2k_{\mathrm{bf}} - 2\left(\varnothing_{\mathrm{Np1}} k_{\mathrm{Np1}} + \varnothing_{\mathrm{Np2}} k_{\mathrm{Np2}}\right) + \varnothing k_{\mathrm{bf}}} \quad (3.1)$$

$$\varnothing = \varnothing_{\mathrm{Np1}} + \varnothing_{\mathrm{Np2}} \quad (3.2)$$

The findings indicated a direct correlation between the thermal conductivity of the nanofluids and the concentration of nanofluid volume. HNFs consist of several nanoparticles combined together. **Equation 3.1** roughly incorporated the variables φ_{Np1} and φ_{Np2}. The relative thermal conductivity of HNFs is significantly influenced by the mixing ratio of nanoparticles. Due to the various nanoparticle combinations that can produce HNF, the formula mentioned earlier cannot be considered a universal formula for estimating the relative thermal conductivity. Researchers consistently formulated their own model equations, tailored to the specific nanoparticle kind and experimental conditions employed.

3.1.1.1 KD2 Pro Thermal Analyzer

Given the lack of theoretical models for HNFs, the researchers employed the KD2 Pro instrument to measure the relative thermal conductivity (kH-Nf) of the HNF. The KD2 Pro thermal property analyzer is a compact and mobile device that runs on batteries and functions by utilizing the transient line heat source method. This device has the ability to measure the thermal conductivity, thermal diffusivity, specific heat (heat capacity), and thermal resistivity of various materials, including low-viscosity liquids and solids. A 6-cm (small) single-needle sensor (KS-1) with a diameter of 1.3 mm and a wire length of 0.8 m is used to test the thermal conductivity of nanofluids. The sensor functions within a range of 0.02 to 2.00 watts per meter Kelvin.

3.1.1.1.1 Operating Procedure

As depicted in **Figure 3.2**, the probe is introduced into the nanofluid sample. The sample quantity is often variable, but a minimum amount of sample is necessary to ensure that at least 75% of the probe is submerged in the provided sample. To activate the power button, press and hold the on button and then select the thermal conductivity measurement option from the Menu button. According to the transient line heat source approach, heat input is applied to increase the temperature of the probe and the surrounding fluid. The KD2 Pro system calculates the thermal conductivity of the fluid by analyzing the combined needle temperature data obtained during both the heating and cooling phases.

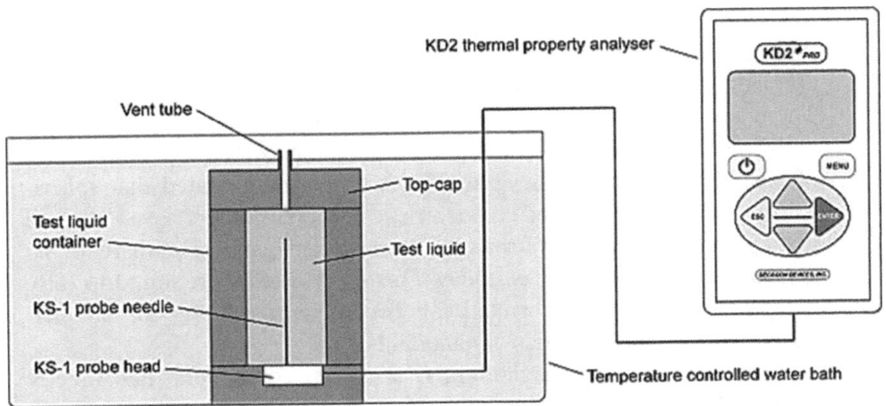

FIGURE 3.2 Kd2 Pro thermal analyzer setup for the measurement of thermal conductivity of nanofluids.[2]

Source: Ebrahimi R, de Faoite D, Finn DP, Stanton KT. Accurate measurement of nanofluid thermal conductivity by use of a polysaccharide stabilizing agent. *Int J Heat Mass Transf.* Pergamon; 2019;136:486–500.

The summary of existing literature on the impact of various parameters on the thermal conductivity of nanofluids is as follows:

i. The thermal conductivity of nanofluids increased in a linear manner with the concentration of nanofluids due to the enhanced Brownian motion of particles.
ii. The shape and structure of the nanoparticle will determine its thermal conductivity. The thermophysical characteristics are contingent upon the nanoparticle material's composition. According to general understanding, it is anticipated that particles composed of materials with high thermal conductivity will result in a high thermal conductivity for the nanofluid, as depicted in **Figure 3.1**. The HNF's overall thermal conductivity will be improved by combining the chemically inert particles with the highly conductive nanoparticles. The improved thermophysical characteristics of HNFs have highlighted the importance of the combined impact of nanoparticles, as opposed to nanofluids with only a single type of particle.
iii. The selection of the base fluid is determined by the specific purpose it will be used for. For instance, in frigid regions, instead of solely utilizing water, a mixture of water and EG is employed. Regardless of the base fluid, a distinct stability mechanism must be adhered to achieve long-term dispersion of nanoparticles. The ability to determine the enhancement rate of nanofluids relies on two crucial parameters: fluid viscosity and thermal conductivity.
iv. The size of nanoparticles is a crucial aspect in enhancing the properties of nanofluid, in addition to the type of nanoparticles used. The size of particles directly affects the Brownian motion, and greater random motion improves the thermal bridging phenomenon. In addition, when particles

are at the nanoscale scale, the gravitational force is nullified, preventing the particles from clumping together. Therefore, one can anticipate a more efficient transfer of the properties.

v. The majority of studies focus on the impact of nanoparticle composition and dimensions, but there is a scarcity of research examining the influence of particle morphology. Several studies have demonstrated that spherical particles have enhanced characteristics, whereas others have indicated that cylindrical or tubular forms are more advantageous. But there are no conclusive proofs made even today. Further research is required to fully comprehend the precise correlation between the morphology of nanoparticles and the characteristics of nanofluids.

vi. The temperature of the nanofluid is a critical component that directly affects the thermal conductivity of nanofluids. As the temperature increases, the nanoparticles absorb heat, leading to an increase in the kinetic energy of the particles. This immediately impacts the Brownian motion of the particles, hence enhancing the thermal conductivity of the nanofluid. This phenomenon exhibits greater intensity at elevated nanofluid concentrations. However, it is important to note that this could potentially lead to the clustering of the particles. Therefore, it is necessary to conduct more comprehensive experiments to determine the optimal operating temperature for the nanofluid.

3.1.2 Viscosity

As rheology is a field dedicated to examining the flow characteristics of fluids, each new set of particles in HNFs is a significant parameter. Depending on whether the fluid falls into the Newtonian or non-Newtonian category, its response to changes in shear rate and duration can be comprehended. Moreover, if the inclusion of nanoparticles significantly alters the viscosity of the fluid, it will have an impact on both pump efficiency and pumping power. Measuring the viscosity of every proposed nanofluid and HNF is crucial.

Nanofluids generally exhibit increased thermal conductivity, density, and viscosity relative to the basic fluid, whereas their specific heat capacity is typically lower. Moreover, heightened viscosity poses a drawback in nanofluid applications as it amplifies the friction factor, leading to a rise in pressure drop and, consequently, an increase in the demand for pumping power. The specific heat capacity of a nanofluid is a crucial thermodynamic characteristic that determines the efficiency of heat transport in a colloidal solution.[3] Multiple studies have demonstrated that nanofluids possess a reduced specific heat capacity compared to base fluids, mostly due to the inherently lower heat capacity of nanoparticles.

There are various factors, such as nanoparticle concentration, temperature, and particle size, which affect the resulting properties of nanofluids. The purpose of this chapter is to provide a comprehensive understanding of the thermophysical and optical characteristics of nanofluids. The overall effect of many factors, such as nanofluid concentration, temperature, and nanoparticle size, on the thermophysical properties of nanofluids is demonstrated in **Table 3.1**.

TABLE 3.1
Impact of Several Variables on the Thermophysical Characteristics of Nanofluids

	List of Variables			
Thermophysical Property	**Nanofluid Concentration**	**Temperature**	**Nanoparticle Size**	**Overall Effect**
Thermal conductivity	Increases	Increases	Contradictory results	Increases
Density	Increases	Decreases	Not available	Increases
Viscosity	Increases	Decreases	Contradictory results	Increases
Specific heat	Decreases	Contradictory results	Increases	Increases
Surface tension	Contradictory results	Decreases	Not available	Increases

Estimating the viscosity of nanofluids, which are composed of solid particles in a fluid media, is challenging due to the complex nature of the theoretical models involved. In the early 1900s, Einstein[4] formulated an equation to describe the viscosity of a low-viscosity fluid when a single particle is suspended in it (**Equation 3.3**). Here, the subscripts *nf* and *bf* signify nanofluid and base fluid, respectively. Furthermore, this equation is applicable solely to diluted solutions of nanofluid with a concentration below 0.02 Vol%. Brinkman et al. in 1952[5] [5] expanded upon the Einstein theory by taking into account the presence of a limited amount of nanofluids (as described by **Equation 3.4**). In addition, Batchelor (1977)[6] incorporated the influence of Brownian motion of the particles in the nanofluid and derived **Equation 3.5**. Nguyen et al.[7] formulated model equations to measure the viscosity of CuO/water and Al_2O_3/water nanofluids at various nanoparticle sizes ranging from 0.15 to 13 Vol%. These equations are derived from a comprehensive analysis of the given equations as shown in **Equation 3.6**. The findings from **Figure 3.3** indicate that the viscosity of the nanofluid increased as the concentration of the nanofluid increased. The viscosity of the nanofluid with a particle size of 47 nm is evidently greater than that of the nanofluid with a particle size of 36 nm. Equation 3.6 is applicable for quantifying the viscosity of Al_2O_3/water nanofluid-containing particles with a size of 47 nm. **Equation 3.7** is applicable for quantifying the viscosity of Al_2O_3/water nanofluid-containing particles with a size of 36 nm. **Equation 3.8** is applicable for quantifying the viscosity of CuO/water nanofluid. Ahmadi et al. 2018[7] developed the model **Equation 3.9** for Fe_3O_4-MWCNT/EG HNF as a function of temperature (25°C to 50°C) and nanofluid concentration with an R^2 value of 0.99.

$$\mu_{\text{nf}} = \mu_{\text{bf}}\left(1+2.5\varphi\right) \tag{3.3}$$

$$\mu_{\text{nf}} = \mu_{\text{bf}}\frac{1}{\left(1-\varphi\right)^{2.5}} \tag{3.4}$$

$$\mu_{\text{nf}} = \mu_{\text{bf}}\left(1+2.5\varphi+6.5\varphi^2\right) \tag{3.5}$$

$$\mu_{\text{nf}} = \mu_{\text{bf}}\left(0.9024e^{0.148\varphi}\right) \tag{3.6}$$

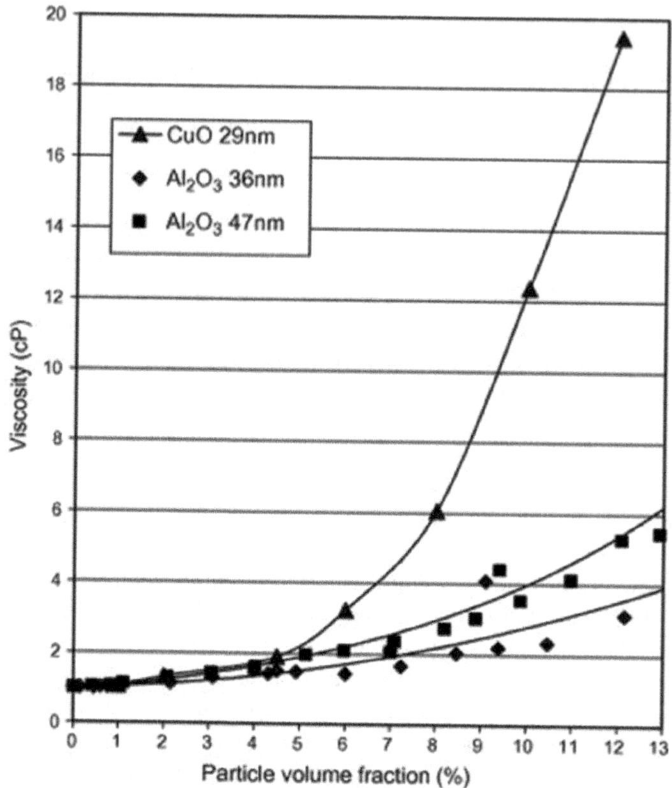

FIGURE 3.3 Effect of nanoparticle type and its size on the viscosity of nanofluids as a function of nanofluid concentration.[8]

Source: Nguyen CT, Desgranges F, Roy G, Galanis N, Maré T, Boucher S, et al. Temperature andparticle-sizedependentviscositydataforwater-basednanofluids—Hysteresisphenomenon. *Int J Heat Fluid Flow*. Elsevier; 2007;28:1492–506.

$$\mu_{\text{nf}} = \mu_{\text{bf}}\left(1+0.025\varphi+0.015\varphi^2\right) \quad (3.7)$$

$$\mu_{\text{nf}} = \mu_{\text{bf}}\left(1.475-0.319\varphi+0.051\varphi^2+0.009\varphi^3\right) \quad (3.8)$$

$$\frac{\mu_{\text{nf}}}{\mu_{\text{bf}}} = \frac{-2.0987+\left(4.65\varphi\right)^{0.0969}+\left(0.8702T\right)^{0.2633}+62323.13\varphi^2}{143.107T^2} \quad (3.9)$$

3.2 OPTICAL PROPERTIES OF NANOFLUIDS

Contrary to the thermophysical properties, the optical properties of typical fluids can be altered by including nanoparticles. Nevertheless, in terms of optical characteristics, a discernible modification may be detected even with a minor fluctuation in nanofluid concentrations (<0.01 Vol%), which is below the threshold for alterations in thermophysical properties. The optical characteristics of solids exhibit significant disparities in comparison to the thermophysical properties of liquids.[9] Fluids are characterized by their optical properties, which are measured using the extinction coefficient (k_e), absorptivity (α), transmittivity (τ), and reflectivity (ρ). Fluids, such as water and ionic liquids, exhibit low levels of attenuation in the visible spectrum, which ranges from approximately 400 to 750 nm. The visible spectrum pertains to a distinct range of wavelengths. When light comes into contact with fluids, some of the incoming light is absorbed while the rest is scattered.[10] The extinction coefficient (k_e) is calculated by considering the absorption and scattering of incident light (I) in relation to this phenomenon. In this context, K_a and K_s denote the absorbance and scattering coefficients, respectively.[11] The understanding of K_e is essential for the comprehension of how incident light is absorbed, especially in the context of solar collector applications.[12]

$$\frac{dI}{dx}+\left(K_a+K_s\right)I=0 \tag{3.10}$$

$$K_a+K_s=K_e \tag{3.11}$$

Black objects possess the capacity to completely assimilate incoming light, which is why the majority of solar collectors are coated with a black hue to optimize the absorption of solar energy. Later researchers also endeavored to employ black fluids including nanoparticles; however, these efforts were not endorsed due to their intrinsic constraints. This hypothesis has been proposed to integrate the use of nanofluid to enhance the absorption of solar light. Nanofluids comprise nanoparticles with optical characteristics that differ from those of the fluids. As a result, the characteristics of the fluids are greatly altered in a way that enhances their combined effect. The extinction coefficients of nanofluids are influenced by various parameters, including the characteristics of the nanoparticles (such as type, shape, size, and qualities like wave and optical path length), as well as the concentration of the nanofluid and the type and concentration of the surfactant. At elevated concentrations of nanofluids, the particles undergo enhanced Brownian motion, resulting in heightened interaction with incident light. As a result, the extinction coefficient also rises. For HNFs, the extinction coefficient was determined by adding together the separate extinction coefficients of the mono nanofluids.[13]

Amir Menbari et al.[14] conducted a study to investigate the extinction coefficient of $CuO–Al_2O_3$/water HNFs. In this study, HNFs were stabilized using the acoustic cavitation principle by continuous ultrasonication. The nanofluid's stability was assessed with a UV spectrometer. In this context, the author has utilized Mie theory to calculate the value of the extinction coefficient, as indicated here.

$$k_e=k_a+k_s=\frac{3\varnothing}{2D_P}\left(Q_{a,\lambda}+Q_{s,\lambda}\right) \tag{3.12}$$

In this context, the symbol φ indicates the concentration of nanofluid, D_P stands for the diameter of the particle, and Q_a and Q_s represent the efficiency of scattering and absorption, respectively. The extinction coefficient of the Al_2O_3/water HNF was nearly equivalent to the sum of the extinction coefficients of the individual mono nanofluids. Moreover, it demonstrated a direct relationship with the increasing levels of nanofluid.

3.3 ELECTRICAL PROPERTIES OF NANOFLUIDS

The electrical conductivity (EC-σ, μS/cm) of a nanofluid is determined by the ability of charged particles in the mixture to transfer their charges to the appropriate electrodes when an electric potential is introduced. The ability in question is related to the particle's capacity to bear electric charges. Electroconductive fluids have been used in several technologies and sectors for magneto-hydrodynamic performances. They have several applications such as field-induced pattern formation in colloidal dispersion, sensors, and electrical applications. Electrically conductive fluids are utilized in these circumstances. Various scientists have recently tested the electrical conductivity of numerous nanoparticles, such as carbon nanotubes, graphene sheets, alumina, silver, silicon carbide, and zinc oxides, in diverse base fluids. Conductivity meters are commonly employed to quantify the electrical conductivity of nanofluids.

The stability of a nanofluid is a crucial determinant in assessing its electrical conductivity. The clustering of nanoparticles leads to a decrease in the number of electric charge carriers, resulting in a decrease in the electrical conductivity of the nanofluid due to a decrease in density. Furthermore, the augmented aggregates contribute to the modification of the Brownian motion of the particles, which is another factor leading to the decrease in conductivity. Furthermore, the electrical conductivity value exhibits a pattern of increase as the concentration of nanofluid increases, the size of nanoparticles decreases, and the temperature of the nanofluid increases. The heightened Brownian motion of the particles is a primary factor contributing to the rise.

In one of the examples, Minea et al. (2012)[15] investigated the electrical conductivity of an alumina/water nanofluid. They examined how the nanofluid concentration and temperature affected the conductivity. The results demonstrated a positive correlation between nanofluid concentration and temperature with an improvement in electrical conductivity. An increase of around 390% was seen while utilizing a 4 Vol% Al_2O_3/water nanofluid at a temperature of 60°C. The electrical conductivity of distilled water increased from 5 to 1903 μS/cm when 4 Vol% of nanofluids were added. The extended duration of stability in nanofluids is a primary factor contributing to the increase in their electrical conductivity. A statistical model has been developed to describe the relationship between electrical conductivity and temperature, as well as the concentration of nanofluid, based on the obtained results (**Equation 3.13**).

$$\begin{aligned}\text{Electrical Conductivity} = {} & 176.69 + 588.41\varphi - 13.64T - 86.31\varphi^2 + 0.36T^2 + 1.07T\varphi \\ & + 11.06\varphi^3 - 0.003T^3 + 0.18T^2\varphi - 1.01T\varphi^2 \end{aligned} \tag{3.13}$$

A recent study conducted by Chereches in 2019[16] experimentally examined the electrical conductivity of mono nanofluids and HNFs consisting of SiO_2, TiO_2, and Al_2O_3 nanoparticles in a base fluid of water. The results indicate that the impact of concentration on enhancing the electrical conductivity of nanofluid is significantly more than the impact of temperature. Both mono and HNFs exhibit greater electrical conductivity than distilled water. Titania nanofluids exhibited higher values, mostly due to the increased electrical conductivity of titania nanoparticles compared to silica nanoparticles, as depicted in **Figures 3.4** and **3.5**.

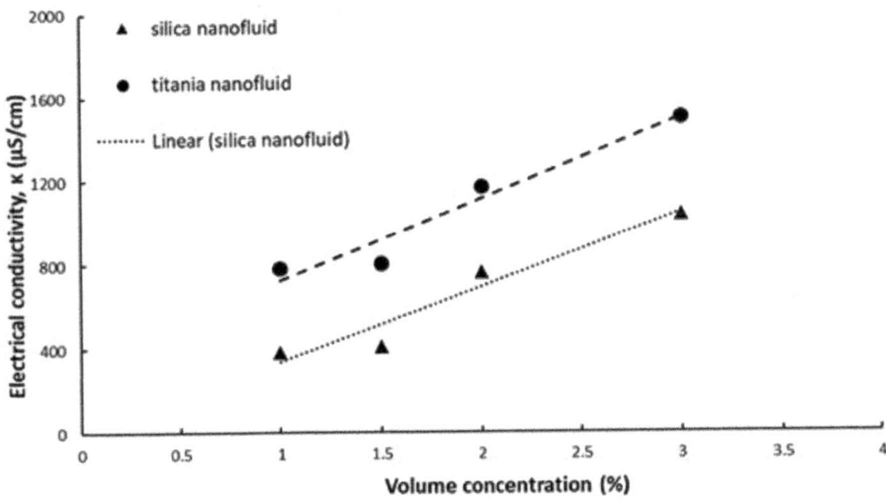

FIGURE 3.4 Effect of nanofluid concentrations on the electrical conductivity of nanofluid.[16]

Source: Chereches EI, Minea AA. Electrical conductivity of new nanoparticle enhanced fluids: An experimental study. *Nanomaterials.* 2019;9: Open access.

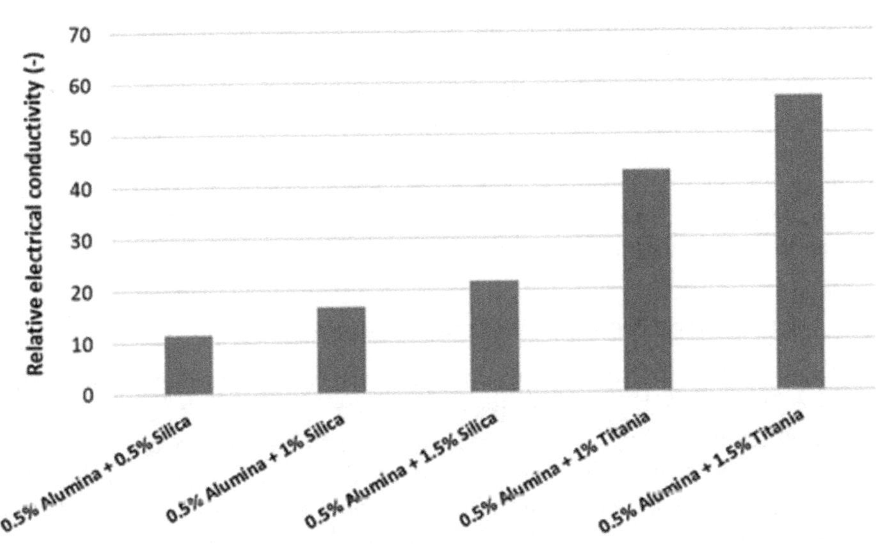

FIGURE 3.5 Effect of nanofluid concentrations on the electrical conductivity of Alumina-Silica hybrid nanofluid.[16]

Source: Chereches EI, Minea AA. Electrical conductivity of new nanoparticle enhanced fluids: An experimental study. *Nanomaterials.* 2019;9: Open access

3.4 MAGNETIC PROPERTIES OF NANOFLUIDS

Magnetic nanofluids consist of magnetic or ferro nanoparticles suspended in a colloidal solution composed of non-magnetic base fluids. In contrast, these intelligent fluids allow for the controlled manipulation of nanoparticle movement by the use of an external magnetic field, enabling unrestricted flow in any direction. Iron (Fe), magnetite (Fe_3O_4), maghemite (Fe_2O_3), and iron nanoparticle alloys, such as core-shell encapsulation or doping with platinum (Pt), copper (Cu), gold (Au), silver (Ag), nickel (Ni), and cobalt (Co) nanoparticles, are all considered general magnetic nanoparticles (MNPs). One can enhance the combined powers of MNPs by adding them to the base fluids using suitable stabilization procedures. By comprehending the ferromagnetic hydrodynamics of these nanofluids, it becomes feasible to manipulate the characteristics of nanoparticles for a diverse range of real-time applications.

3.4.1 Preparation of Magnetic Nanofluids

In general, MNPs exhibit hydrophilic properties. To provide the appropriate stabilization of the colloidal suspension of nanoparticles, the magnetic fluids are coated with surfactants. Surfactants possess a hydrophilic head and a hydrophobic tail. When surfactants are added to ferrofluids, they enhance the steric repulsion between the MNPs, hence improving the stability of nanofluids. Oleic acid (OA) is a commonly employed surfactant for stabilizing MNPs in a base fluid. The carboxylic acid group (head) of OA will bind to the surface of the nanoparticle, while the octadecyl functional group chain (tail) will stabilize the dispersion of the nanoparticle. The use of oleic acid coatings is essential for dispersing MNPs in hydrocarbon-based fluids. These coatings have been the subject of substantial recent research. On the contrary, dodecyl-benzenesulfonic acid surfactants are employed to disperse nanoparticles in water and other organic solvents. The presence of a dual surfactant layer, together with steric repulsion, guarantees the stabilization of the magnetic nanofluid (**Figure 3.6**). A diverse range of magnetic nanofluids are created by employing the same method, varying the base fluids.

3.4.2 Characterization of Magnetic Nanofluids

The magnetic characteristics of magnetic nanofluids can be determined using a magnetometer. By examining the hysteresis profile, the properties of the MNP can be studied. When the hysteresis profile has a reverse pattern, the nanoparticles exhibit superparamagnetic properties, as depicted in **Figure 3.7**. Ferromagnetic MNP exhibits high and irreversible magnetization values. Paramagnetic materials have a feeble affinity for external magnetic fields and do not maintain any magnetization after the removal of the external magnetic field. Examples of such materials include aluminum and platinum. To enhance magnetism, the concept of super magnetization is employed, wherein the particle size is specifically within the nanometer range, namely less than 50 nm. As the particle size decreases, the magnetic particles undergo changes in their overall characteristics, increasing magnetic susceptibility. Ferromagnetic materials, such as iron (Fe) and nickel (Ni), maintain their magnetic characteristics even after the magnetic field is removed.

FIGURE 3.6 Suspension of magnetic nanoparticles (MNPs) in different base fluids. (a) Uncoated MNPs in CCl_4 base fluid. (b) Organic acid (OA)-coated MNPs in CCl_4 base fluid. (c) Dodecylbenzenesulfonic acid (DBSA)-coated MNP in CCl_4 base fluid. (d) Uncoated MNP in water. (E) OA-coated MNP in water. (f) DBSA-coated MNP in water[17]—open access.

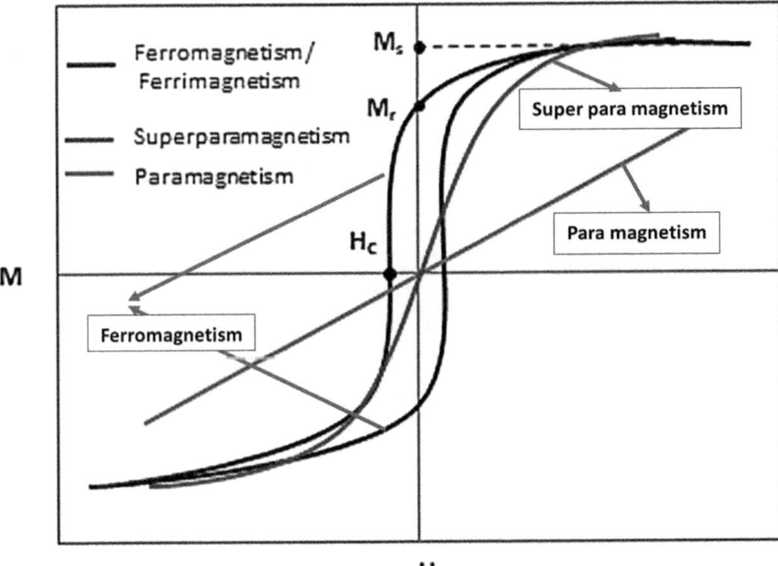

FIGURE 3.7 Magnetic properties of material under magnetization: hysteresis curves.

3.4.3 Biomedical Applications of Magnetic Nanofluids

Due to their exceptional capabilities, MNPs can be controlled using an external magnetic field. This characteristic has led to its use in numerous real-time systems for various biomedical applications.

3.4.3.1 Target Drug Delivery

Traditional delivery methods have several drawbacks, such as a short half-life of the drug, the requirement for a large volume of distribution, the inability to target particular locations for drug administration, a low therapeutic index, poor solubility and stability of the drug, high toxicity, and low patient compliance (**Figure 3.8**). Controlled drug distribution is an effective approach to conventional drug delivery limitations where drugs are carried to a specific location rather than the whole body. In targeted drug delivery, drugs are placed into magnetic nanofluids, which can be carried to specific cells when an external magnetic field is applied. This can significantly decrease the quantity of medication administered through the traditional drug delivery system and minimize the harmful impact on healthy cells.

3.4.3.2 Hypothermia

Hyperthermia refers to the targeted destruction of cancer cells through the use of localized heat. Due to their elevated energy consumption, cancer cells are more vulnerable to heat, which increases their susceptibility. The absorption and release of

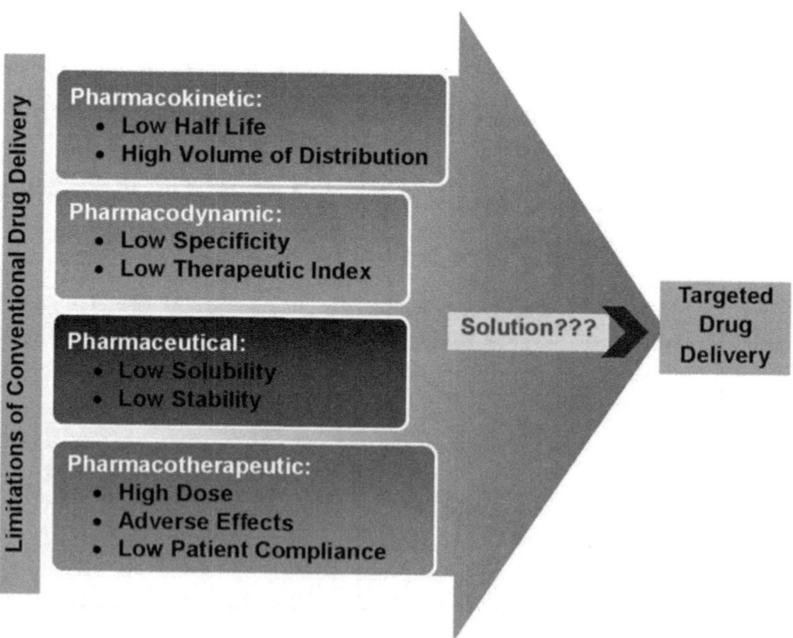

FIGURE 3.8 Limitations of the traditional drug delivery system.[18]

heat by MNPs depend on their size and the characteristics of the magnetic field. The literature survey indicates that the optimal nanoparticle size should not exceed 20 nm and generally falls within the range of 10 to 100 nm. Magnetic nanofluids are being extensively studied for their biocompatibility and low toxicity, making them suitable for hypothermia applications.

Generally, water and hydrocarbons are used as base fluids to disperse the MNPs of size between 20 and 100 nm. The utilization of magnetic nanofluids was initially established in 1957, wherein nanoparticles of ferric oxide (Fe_2O_3) were subjected to heating using a 1.2 MHz alternating current magnetic field, resulting in temperatures ranging from 42°C to 46°C. This approach has been widely used for the treatment of various tumors for a significant period of time. Magnetic nanofluids are being widely researched at various stages of clinical and animal-based testing connected to cancer. It has been demonstrated to be an alternative source with fewer adverse effects compared to surgery, chemotherapy, and radiotherapy.

3.4.3.3 Photothermal Therapy

The application of a light source in conjunction with a localized heating therapy is what is known as photothermal therapy. One example of this is the utilization of nanoparticles that possess a high capacity for light absorption as thermal therapeutic agents. Nanoparticles such as gold, graphene, and carbon nanotubes are frequently utilized in this application; however, one of the obstacles that were encountered was the aggregation of nanoparticles in an unsuitable manner in the cancer location. Consequently, because of the poor dispersion, these therapeutic agents were unable to deliver the heat to places that were not particular. Due to the fact that MNPs are simple to control using an external current magnetic field, they are able to efficiently surround the tumor that is being targeted. Magnetic nanofluids have been demonstrated to be one of the most effective photothermal therapeutic agents due to the features they possess.

3.5 SUMMARY

This chapter addressed the importance of diverse properties of nanofluids and HNFs for prospective real-time applications. The characteristics of nanofluids encompass thermal conductivity, dynamic viscosity, extinction coefficient, electrical conductivity, and magnetic properties. When nanoparticles are in the colloidal form, they exhibit continuous Brownian motion, which synergistically enhances the properties of base fluids. HNFs are composed of several nanoparticles, making them significantly more sophisticated than mono nanofluids. The thermal conductivity of nano fluids increases with both the concentration of nanofluids and the temperature. When dealing with HNFs, the most important factors to consider are the type, compatibility, and mixing ratio of the nanoparticles, as these fluids consist of a combination of many nanoparticles. As the concentration of nanofluid increased, the viscosity exhibited an increase, whereas it dropped with an increase in temperature. The extinction coefficient is one of the important optical properties which represents the absorption and scattering of incident light (I), which are highly significant for understanding how incident light is absorbed, especially in the context of solar collector applications.

The chapter focuses on the comprehensive study of magnetic nanofluids and their prominent role in various biomedical applications. The stability of nanofluids is a crucial factor for their effective application. The correct dispersion of nanoparticles in a base fluid is the limiting element which is restricting industrial applications.

REFERENCES

1. Ukueje, W. E., Abam, F. I., & Obi, A. (2022). A perspective review on thermal conductivity of hybrid nanofluids and their application in automobile radiator cooling. *Journal of Nanotechnology, 1*, 2187932.
2. Ebrahimi, R., de Faoite, D., Finn, D. P., & Stanton, K. T. (2019). Accurate measurement of nanofluid thermal conductivity by use of a polysaccharide stabilising agent. *International Journal of Heat and Mass Transfer, 136*, 486–500.
3. Olia, H., Torabi, M., Bahiraei, M., Ahmadi, M. H., Goodarzi, M., & Safaei, M. R. (2019). Application of nanofluids in thermal performance enhancement of parabolic trough solar collector: State-of-the-art. *Applied Sciences, 9*(3), 463.
4. Einstein, A. (1906). Eine neue Bestimmung der Moleküldimensionen. *Annals of Physics, 324*(2), 289–306.
5. Brinkman, H. C. (1952). The viscosity of concentrated suspensions and solutions. *The Journal of Chemical Physics, 20*, 571.
6. Batchelor, G. K. (1977). The effect of Brownian motion on the bulk stress in a suspension of spherical particles. *Journal of Fluid Mechanics, 83*(1), 97–117.
7. Ahmadi, A., Eshgarf, H., & Afrand, M. (2018). Measuring the viscosity of Fe_3O_4-MWCNTs/EG hybrid nanofluid for evaluation of thermal efficiency: Newtonian and non-Newtonian behavior. *Journal of Molecular Liquids, 253*, 169–177. https://doi.org/10.1016/j.molliq.2018.01.012
8. Nguyen, C. T., Desgranges, F., Roy, G., Galanis, N., Maré, T., Boucher, S., et al. (2007). Temperature and particle-size dependent viscosity data for water-based nanofluids—Hysteresis phenomenon. *International Journal of Heat and Fluid Flow, 28*, 1492–1506.
9. Rabbi, H. M. F., Sahin, A. Z., Yilbas, B. S., & Al-Sharafi, A. (2021). Methods for the determination of nanofluid optical properties: A review. *International Journal of Thermophysics, 42*, article no. 9. https://doi.org/10.1007/s10765-020-02762-0
10. Traciak, J., Sobczak, J., Kuzioła, R., Wasąg, J., & Żyła, G. (2022). Surface and optical properties of ethylene glycol-based nanofluids containing silicon dioxide nanoparticles: An experimental study. *Journal of Thermal Analysis and Calorimetry, 147*, 7665–7673. https://doi.org/10.1007/s10973-021-11067-9
11. Mahian, O., Bellos, E., Markides, C. N., Taylor, R. A., Alagumalai, A., Yang, L., et al. (2021). Recent advances in using nanofluids in renewable energy systems and the environmental implications of their uptake. *Nano Energy, 86*, 106069.
12. Ahmad, S. H. A., Saidur, R., Mahbubul, I. M., & Al-Sulaiman, F. A. (2017). Optical properties of various nanofluids used in solar collector: A review. *Renewable and Sustainable Energy Reviews, 73*, 1014–1030.
13. Menbari, A., Alemrajabi, A. A., & Ghayeb, Y. (2016). Experimental investigation of stability and extinction coefficient of Al2O3–CuO binary nanoparticles dispersed in ethylene glycol–water mixture for low-temperature direct absorption solar collectors. *Energy Conversion and Management, 108*, 501–510.
14. Menbari, A., Alemrajabi, A. A., & Ghayeb, Y. (2016). Investigation on the stability, viscosity and extinction coefficient of CuO–Al2O3/Water binary mixture nanofluid. *Experimental Thermal and Fluid Science, 74*, 122–129.

15. Minea, A. A., & Luciu, R. S. (2012). Investigations on electrical conductivity of stabilized water based Al 2O3 nanofluids. *Microfluidics and Nanofluidics*, *13*, 977–985.
16. Chereches, E. I., & Minea, A. A. (2019). Electrical conductivity of new nanoparticle enhanced fluids: An experimental study. *Nanomaterials*, *9*(9), 1228.
17. Li, L., Li, D., & Zhang, Z. (2022). Colloidal stability of magnetite nanoparticles coated by oleic acid and 3-(N,N-Dimethylmyristylammonio)propanesulfonate in solvents. *Frontiers in Materials*, *9*, 1–10.
18. Tewabe, A., Abate, A., Tamrie, M., Seyfu, A., & Siraj, E. A. (2021). Targeted drug delivery—from magic bullet to nanomedicine: Principles, challenges, and future perspectives. *Journal of Multidisciplinary Healthcare*, *14*, 1711–1724.

4 The Effect of Nanoparticle Mixing Ratio on the Properties of Hybrid Nanofluids

Nanofluids consist of nanoparticles suspended in a base fluid. Ensuring the nanoparticles remain well dispersed is crucial for achieving the desired improvement in characteristics. Considering the advanced nature of hybrid nanofluids (HNFs), it is crucial to take into account the compatibility between the nanoparticles. Recently, several nanoparticles such as Al_2O_3, CuO, ZnO, SiO_2, CNTs (multi-walled carbon nanotubes (MWCNTs) and single-walled carbon nanotubes (SWCNTs)), graphene, Fe_2O_3, Fe_3O_4, TiO_2, and SiC have been incorporated in diverse combinations and distributed in base fluids such as water, motor oil, ethylene glycol, and propylene glycol solutions.

FACTORS AFFECTING THE PROPERTIES OF HNFS

The production of nanofluids involves the straightforward procedure of dispersion of nanoparticles in a base fluid. To utilize nanofluids effectively in real-time applications, it is crucial to comprehend the appropriate stability mechanism of these fluids. The literature review indicates that the stability of HNF is dependent on the following:

- Type and compatibility of nanoparticles
- Optimum mixing ratio of the nanoparticles
- Optimum concentration of nanofluid
- Selection of appropriate base fluid
- Type of surfactant

4.1 TYPE AND COMPATIBILITY OF NANOPARTICLES

The ability to achieve a synergistic impact by combining different nanoparticles depends entirely on the properties of the nanoparticles and the method used to prepare the HNF. The crucial inquiry at hand is how to select the nanoparticle. There is a lack of data to support the specific rationale for selecting the particular combination of nanoparticles in their research. One possible explanation is that the selection of hybrid nanoparticles is determined by the desired outcome and the specific application in which they will be used. **Table 4.1** displays several hybrid

DOI: 10.1201/9781003595137-4

TABLE 4.1
Different Kinds of Hybrid Nanoparticles Explored by Well-Known Researchers

Metal Oxide + Metal	Metal Oxide + Metal Oxide	Metal + Non-Metal	Metal Oxide + Non-Metal	Metal + Metal
Al_2O_3/Cu	TiO_2/SiO_2	Ag/GNPs	Fe_2O_3/CNTs	Cu/Zn
MgO/Ag	CuO/ZnO	Ag/MWCNTs	ZnO/MWCNTs	Al/Zn
TiO_2/Ag	TiO_2-CuO/C	Pt/GNPs	TiO_2/CNTs	
TiO_2/Cu	Al_2O_3/CuO	Ni/ND	MgO/MWCNTs	
ZnO/Ag	MgO/ZnO	Ag/Si	GO/MWCNTs	
WO_3/Ag	SiO_2/CuO		SiO_2/GNPs	
	Al_2O_3/ZnO		TiO_2/MWCNTs	
	Al_2O_3/SiO_2		Fe_3O_4/MWCNTs	
			Al_2O_3/MWCNTs	
			ZnO/DWCNTs	
			Al_2O_3/graphene	
			Co_3O_4/GO	
			Al_2O_3/CNTs	
			SiO_2/MWCNTs	
			TiO_2/SiC	
			Co_3O_4/ND	

nanoparticle combinations investigated by renowned experts up until 2024. These nanoparticles can consist of various types of elements, including metals, nonmetals, and metal oxides. An effective technique for preparing HNF involves the dispersion of nanoparticles of certain components or composite nanoparticles in a base fluid at a precise ratio. A customized sonication process is used to provide a uniform and steady distribution. HNFs can be synthesized using two separate methods.

4.2 OPTIMUM MIXING RATIO OF THE NANOPARTICLES

4.2.1 Case Study 1: CuO-MWCNT/Water HNF For Optimum Mixing Ratio

The cost analysis of the nanofluid is also a vital factor to consider when selecting the nanoparticles. For example, the thermal conductivity of MWCNT is 3000 W/mK, while copper oxide (CuO) has a thermal conductivity of 76 W/mK. The price of multi-walled nanoparticles, with a diameter ranging from 50 to 90 nm, is Rs. 39,460. The price of 25 gm of CuO nanoparticles with a diameter of 50 nm from the Sigma-Aldrich database is Rs. 11,220. CuO particles with a size less than 10 micrometers are priced at around Rs. 5050. Furthermore, micron-sized particles have a tendency to settle rather than remain in suspension. It is possible for the flow of liquid through a microchannel to cause pipe blockage. Therefore, we must go for the nanoparticle instead of the micron-sized particle.

It is vital to adjust the mixing ratio of the nanoparticles to ensure that the MWCNT-CuO-based HNF can be utilized in any of the industrial applications. To make use of the HNF as a coolant, the objective of the optimization process should be to increase the thermal conductivity of the nanofluid while simultaneously reducing the expenses associated with the preparation of the HNF. An estimate of the costs involved in generating 10 liters of an HNF containing 0.01 volume percent of CuO-MWCNT and water is presented in **Table 4.2**. When it comes to binary base fluids, the same table can also be utilized to determine the quantity of nanoparticles that should be introduced. In this particular instance, the mixture of water and ethylene glycol is utilized. Alterations to the values in the table are possible depending on the type of the base fluid. Considering that we are simply using water as the base fluid in this instance, the third column of **Table 4.3** is assumed to be zero. The following equation is used to construct **Table 4.3.** Modifications to this formula are possible according to the nanoparticles that are chosen.

$$\text{Solid volume fraction of hybrid nanofluid}(\varnothing) = \left(\frac{\left[\frac{W}{\rho}\right]_{\text{CuO}} + \left[\frac{W}{\rho}\right]_{\text{MWCNT}}}{\left[\frac{W}{\rho}\right]_{\text{CuO}} + \left[\frac{W}{\rho}\right]_{\text{MWCNT}} + \left[\frac{W}{\rho}\right]_{\text{water}}} \right) \times 100 \tag{4.1}$$

Here, W and ρ represent the weight and density of nanoparticles.

Properties of nanoparticles are as follows:

Density (ρ):	MWCNT:	2.1 gm/cm^3
	CuO:	6.31 gm/cm^3
Thermal conductivity:	Water:	0.6 W/mK
	CuO:	76 W/mK
	MWCNT:	3000 W/mK

After the preparation of the nanofluids, a sequential numbering system is assigned, ranging from 1 to 11. **Table 4.3** shows that serial numbers 1 and 11 correspond to mono nanoparticle-based nanofluids, specifically MWCNT/water and CuO/water. S. No. 2 to 10 in the table represent different combinations of CuO-MWCNT/water HNFs. Once the nanofluids are created, their particle size in the dispersion form, zetapotential (measured in millivolts), density (represented by ρ), and thermal conductivity (shown by k) are measured. The obtained values are then organized and presented in **Table 4.4**. Due to the varying diameters of the nanoparticles, the resulting HNF particles likewise exhibited various sizes (refer to **Table 4.4, Column 4**). Zetapotential refers to the net electric charge on the surface of a particle when it is scattered in a liquid medium. It quantifies the extent of repulsion between nanoparticles. Various methods exist for quantifying the zetapotential of nanofluids, and specialized equipment is also accessible that can provide direct measurements of this value. The zetapotential value (ζ) of a nanofluid is directly proportional to the repulsion forces present within the fluid. Therefore, a higher zetapotential value indicates stronger repulsion forces and greater stability of the nanofluid.

TABLE 4.2

Sample Calculation Table for Producing 0.01 Vol% of CuO-MWCNT/Water-Based HNF—Cost Estimation

S. No.	Water (%)	EG (%)	Φ (Vol%)	Mixing Ratio (%)		Water (mL)	EG (mL)	Nanoparticle Volume	MWCNT Volume	MWCNT (gm)	CuO Volume	CuO (gm)	Cost of the HNF (10 Liters)
				CuO	MWCNT								
1	100	0	0.01	0	100	9999	0	1	1	2.1	0	0	3314
2	100	0	0.01	10	90	9999	0	1	0.9	1.89	0.1	0.631	3265.7
3	100	0	0.01	20	80	9999	0	1	0.8	1.68	0.2	1.262	3217.7
4	100	0	0.01	30	70	9999	0	1	0.7	1.47	0.3	1.893	3169.6
5	100	0	0.01	40	60	9999	0	1	0.6	1.26	0.4	2.524	3121.6
6	100	0	0.01	50	50	9999	0	1	0.5	1.05	0.5	3.155	3073.5
7	100	0	0.01	60	40	9999	0	1	0.4	0.84	0.6	3.786	3025.4
8	100	0	0.01	70	30	9999	0	1	0.3	0.63	0.7	4.417	2977.4
9	100	0	0.01	80	20	9999	0	1	0.2	0.42	0.8	5.048	2929.3
10	100	0	0.01	90	10	9999	0	1	0.1	0.21	0.9	5.679	2881.3
11	100	0	0.01	100	0	9999	0	1	0	0	1	6.31	2833.2

CuO, copper oxide; EG, ethylene glycol; HNF, hybrid nanofluid; MWCNT, multi-walled carbon nanotube nanoparticle; Vol%, volume percentage.

TABLE 4.3
Optimization of Nanoparticle Mixing Ratios of 0.01 Vol% in a 100-mL Water-Based Hybrid Nanofluid

S. No	Mixing Ratio CuO	Mixing Ratio MWCNT	Size (nm)	Zetapotential (mV)	Density (kg/m^3)	Thermal Conductivity (k) W/mK	Cost of the HNF (10 liters)
1	0	100	20	41	20	1.31	3314
2	10	90	25	45	25	1.27	3266
3	20	80	28	42	28	1.42	3218
4	30	70	25	39	25	1.38	3170
5	40	60	32	40	32	1.41	3122
6	50	50	35	49	35	1.39	3073
7	60	40	28	42	28	1.40	3025
8	70	30	25	48	25	1.42	2977
9	80	20	**22**	**51**	**22**	**1.52**	**2929**
10	90	10	28	49	28	1.24	2881
11	100	0	30	48	30	1.09	2833

4.2.2 Case Study 2: Ternary HNF for Thermal Management Systems

The performance of an HNF that was based on ternary nanoparticles was evaluated by Xuan et al.,[1] who also emphasized the significance of the nanoparticle mixing ratio in relation to the thermophysical properties of the nanofluid through the utilization of sensitivity analysis and regression analysis methodologies. For this investigation, three different nanoparticles were utilized: alumina (Al_2O_3), copper (Cu), and copper oxide (CuO). It has been determined that the nanoparticles have a thermal conductivity of 40, 400, and 77 W/mK, respectively. Utilizing **Equation 4.2**, the required quantity of nanoparticles to be added has been determined based on the proposed concentration of the nanofluid. The variables W and ρ, which reflect the weight and density of the nanoparticles, have been utilized to compute the amount that is needed. Additionally, the density of the HNF was determined using Equation 4.3, where the subscripts NP and BF, which stand for nanoparticle and base fluid, respectively, were used to estimate the density. When doing the regression analysis, the thermal conductivity and viscosity of the nanofluid are determined as a function of the concentration of the nanofluid and the temperature.

$$\text{Solid volume fraction of hybrid nanofluid}(\varnothing) = \left(\frac{\left[\frac{W}{\rho}\right]_{Al_2O_3} + \left[\frac{W}{\rho}\right]_{Cu} + \left[\frac{W}{\rho}\right]_{CuO}}{\left[\frac{W}{\rho}\right]_{Al_2O_3} + \left[\frac{W}{\rho}\right]_{Cu} + \left[\frac{W}{\rho}\right]_{CuO} + \left[\frac{W}{\rho}\right]_{water}} \right) \times 100 \quad (4.2)$$

$$\text{Density of hybrid nanofluid}(\rho_{HNF}) = (\varnothing_{NP1}\rho_{NP1} + \varnothing_{NP2}\rho_{NP2} + \varnothing_{NP3}\rho_{NP3}) + [1 - \varnothing_{NP1} - \varnothing_{NP2} - \varnothing_{NP3}]\rho_{BF} \quad (4.3)$$

Results showed that the viscosity of nanofluid was increasing along with increasing nanofluid concentrations, and, with the reduced intermolecular forces, viscosity decreased when temperature was increased from 20°C to 60°C. Moreover, with the increased kinetic energy between the particles, Brownian motion was enhanced, and the thermal conductivity of ternary HNFs was enhanced along with temperature. In the experimentation, any one of the nanoparticle's volume fraction was fixed at 20%, and others were varied between 20:60, 30:50, 40:40, 50:30, and 60:20. Like this around 15 sets of experiments were conducted. Lower viscosity and higher thermal conductivity were found at a mixing ratio of 20:50:30 for Al_2O_3–Cu–CuO/W HNF, respectively.

4.3 EFFECT OF PH OF THE SOLUTION ON THE PERFORMANCE OF HNF

The nanoparticles are disseminated in the base fluid, which allows the particle to acquire the surface charge that is present on the particle. In addition to this, the surface charge is entirely dependent on the pH of the substrate fluid. In the event that the pH of the base fluid is changed, the charge that is exerted on the particles will likewise be adjusted. Because of this, the pH of the solution has a direct impact on the way in which the particles interact with one another, which in turn causes the stability of the dispersion to shift. **Figure 4.1** illustrates how the zetapotential of the nanofluid is affected by the pH of the solution. In addition, the value of the zetapotential might be either positive or negative, depending on the surface charge that the nanoparticle has acquired. In light of this, the zetapotential value will shift from positive to negative in

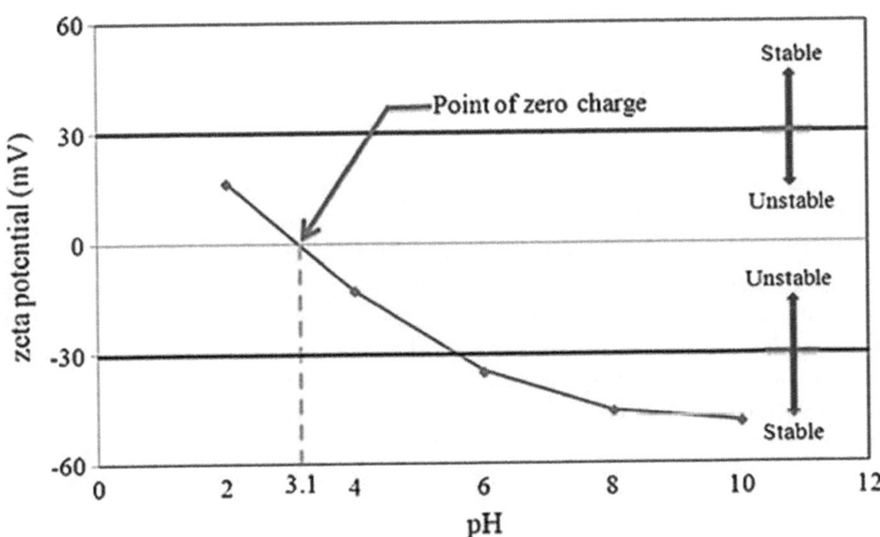

FIGURE 4.1 Effect of pH of the solution on the zetapotential of nanofluid.[2]

Source: Mehrali, M., Sadeghinezhad, E., Latibari, S.T. et al. Investigation of thermal conductivity and rheological properties of nanofluids containing graphene nanoplatelets. *Nanoscale Res Lett.* **9**, 15 (2014). https://doi.org/10.1186/1556-276X-9-15, 5. Open access.

tandem with the change in the pH of the surrounding solution. The value of the zetapotential is positive when the pH is low and negative when the pH is high. The term "isoelectric point" (IEP) refers to the point at which the zetapotential shifts from a positive to a negative value. At the IEP, the zetapotential value is zero, which is equal to zero millivolts and the nanofluid is the least stable at this point due to the strong attraction that exists between the particles. Maintaining the pH of the nanofluid in a manner that is insensitive to IEP will, thus, increase the repulsions that exist between the particles, thereby enhancing the stability of the dispersion. In general, zetapotential values that are higher than 30 mV are considered to be conducive to stability.

In accordance with **Table 4.1**, the HNF composed of CuO-MWCNT and water exhibited zetapotential values that were larger than 30 mV across all mixing ratios. This indicates that the HNF exhibited a longer stability.

Important Point: The pH of the solution can be changed in the experiment by adding HCl or NaOH solutions, and then the zetapotential values are measured after 10 to 15 minutes of ultrasonication.

Even though the density of the HNF changed due to the varying mixing ratios of the nanoparticles, the values were quite comparable to the density of water during the experiment. Due to the extremely low concentration of the nanofluid, there was not a significant amount of change observed in the density. This demonstrated that there is no additional burden associated with pumping the nanofluid in comparison to pumping water. Last but not least, to make use of the CuO-MWCNT/water combination in a heat transfer application, the thermal conductivity of the mixture must be higher than that of the starting fluid. The base fluid in this case is water, which has a thermal conductivity value of 0.61 W/mK. The addition of CuO-MWCNT nanoparticles at a mixing ratio of 80:20 has resulted in a rise in the thermal conductivity of the base fluid to 1.52 W/mK. What this means is that the value of thermal conductivity has increased by 153%.

The enhancement in the thermal conductivity of nanofluids can be attributed mostly to the Brownian motion of nanoparticles. The motion induced the particles to assimilate heat from the surrounding base fluid, hence enhancing the thermal conductivity of nanofluids. Moreover, the combined effect of hybrid nanoparticles resulted in a greater increase in thermal conductivity compared to using only one type of nanoparticle, when dissimilar nanoparticles were suspended in the base fluid. Furthermore, 10 liters of nanofluid costs approximately Rs 2929, which is way lower than the mono nanofluids of CuO and MWCNT. This allows for simple adjustments to the cost of the HNF, increasing the system's cost-effectiveness. Although increasing the concentration of nanoparticles in a mono nanofluid makes it more susceptible to cluster formation, it is recommended to include one extra nanoparticle with improved thermal conductivity. Implementing this measure will reduce the chance of clustering and the consequent increase in costs. This proves that HNFs outperform mono nanofluids.

4.4 OPTIMUM CONCENTRATION OF NANOFLUID

After the precise type of nanoparticles has been identified, the optimization of nanofluid concentrations becomes an important component that has a direct influence on the performance of the HNF. The thermophysical properties of the base fluid are directly improved in a linear fashion when the nanoparticles are increased with increasing concentrations. In the case of an application involving heat transfer, the nanoparticles that were added served as thermal bridges and shifted the method of heat transmission from convection to conduction. Furthermore, when the Brownian motion was complete, the nanoparticles boosted the surface area of the heat transfer and the overall performance of the heat transfer.

On the contrary, choosing the optimal value is an important topic that needs to be discussed, owing to the fact that it has a direct influence on the HNF's performance as well as its cost. Furthermore, when there is a high concentration of particles, the attractive interactions that occur between them may be able to outweigh the repulsive forces, which results in the particles clumping together. As the size of the particles grows, there is a larger tendency for the particles to settle at the bottom, and there is also a noticeable increase in the thermal resistance. When it comes to this topic, the vast majority of studies concentrate primarily on nanofluids that have a solid volume fraction that is lower than 2 Vol%. Furthermore, lowering the concentration of nanofluid results in a reduction in the degree to which the process's costs vary. This enables the flexibility to simply adjust the cost of the HNF, hence enhancing the cost-effectiveness of the system. Although raising the concentration of nanoparticles in a mono nanofluid makes it more prone to cluster formation, it is advisable to incorporate an extra nanoparticle with superior thermal conductivity. Implementing this measure will decrease the likelihood of clustering and the subsequent rise in expenses. This demonstrates how HNFs outperform mono nanofluids.

4.4.1 Case Study 3: Utilization of MWCNT/Water and Ag-MgO/Water Nanofluids for Increasing the Efficiency of Photovoltaic Thermal System

The collecting of solar energy is a sustainable and renewable energy source that is becoming increasingly important due to the rising need for electricity. Solar collectors and photovoltaic panels are commonly used for heating and generating electricity in real-time applications, with respective efficiencies of 75% and 20%. **Figure 4.2** demonstrates the utilization of PV cells for power generation, which is then stored in a battery for usage in residential heating systems. The unutilized solar energy in the photovoltaic panel causes heating in a specific area, resulting in a reduction in the efficiency of the panel cells. As the driving force for heat transfer diminishes, the effectiveness of the solar panel also decreases. To overcome this constraint, one can circumvent it by efficiently harnessing and transforming all the radiation into electrical energy, instead of only raising the temperature of the panel. Considerable effort has been focused on improving the efficiency of power conversion in this field, leading to the development of photovoltaic thermal (PVT) systems.

Conversely, PVT systems consist of multiple solar thermal collector systems and two solar panels, which are capable of generating both heat and power. Thermal carrier fluids are given special importance in PVT systems due to this reason. Initially, several research studies were conducted using basic fluids such as water. However, the limited thermal conductivity of water resulted in an insignificant improvement in thermal efficiency. Nanofluids are employed as both carrier fluids and coolants

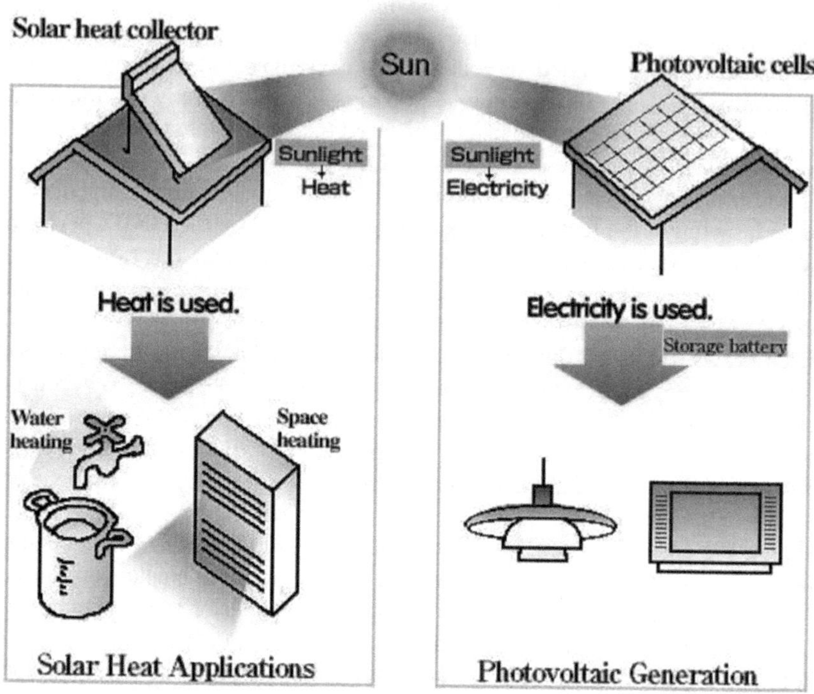

FIGURE 4.2 Difference between solar collector and photo voltaic system.

because nanoparticles possess a greater thermal conductivity compared to conventional liquids, making them more effective in heat transfer. Several studies have employed nanofluids as the working fluid in the PVT system. All of these investigations have indicated that the system's performance is markedly enhanced in comparison to that of conventional fluid-based coolants.

Moghaddam et al.[3] conducted an experimental investigation in which they investigated the possibility of boosting the efficiency of PVT systems by utilizing nanofluids composed of MWCNT and water as well as Ag-MgO and water. The most essential thing that has been explained is the distinction between the effectiveness of mono and HNF-based coolants for the cooling of photovoltaic panels. Both the concentration and the flow rate of the nanofluid and HNF were adjusted anywhere from 0.5 to 2 Vol% throughout the experiment. At the same concentrations, the thermal conductivity enhancement for MWCNT/water nanofluid is greater than that of Ag-MgO/water HNF. This is because the thermal conductivity of MWCNT is significantly higher than that of Ag and MgO nanoparticles. Furthermore, as demonstrated in **Figure 4.3**, the amount of cooling that was produced by PV panels was reduced by a greater amount of 25% when MWCNT/water nanofluid was utilized as opposed to the Ag-MgO/water HNF at a concentration of 2 Vol%. Along with the improvement in thermal efficiency, there was also a nine percent increase in the efficiency of the electrical conversion.

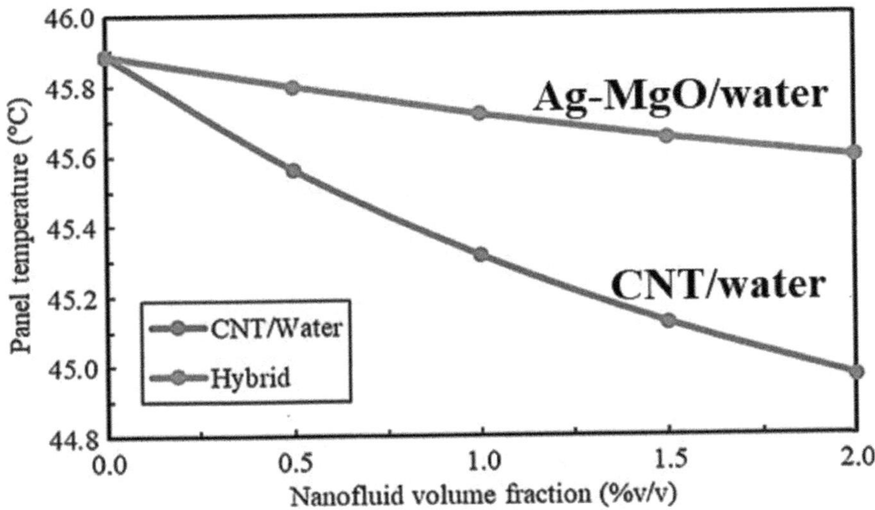

FIGURE 4.3 Effect of multi-walled carbon nanotube (MWCNT)/water nanofluid and Ag-MgO/water hybrid nanofluid concentration on the photovoltaic panel cooling efficiency.[3]

Source: Hormozi Moghaddam, Mohsen, and Maryam Karami. Heat transfer and pressure drop through mono and hybrid nanofluid-based photovoltaic-thermal systems. *Energy Sci Eng.* 10, no. 3 (2022): 918-931. Open access-3

4.5 SELECTION OF APPROPRIATE BASE FLUID

The choice of an adequate base fluid is a critical determinant of the long-term stability of a nanofluid. The primary objective of producing nanofluids is to enhance their thermophysical properties, and the characteristics of the base fluids play a crucial role in achieving this objective.[4] These are the various primary fluids typically employed in the creation of nanofluids:

1. Distilled water
2. Transformer oil
3. Ethylene glycol
4. Propylene glycol
5. Engine oil
6. Kerosene
7. Ethylene glycol–water mixture (50:50, 60:40, 40:60, 25:75, and 75:25)
8. Paraffin oil
9. Gear oil
10. Coconut oil
11. Methanol
12. Ethanol
13. Acetone
14. Motor oil

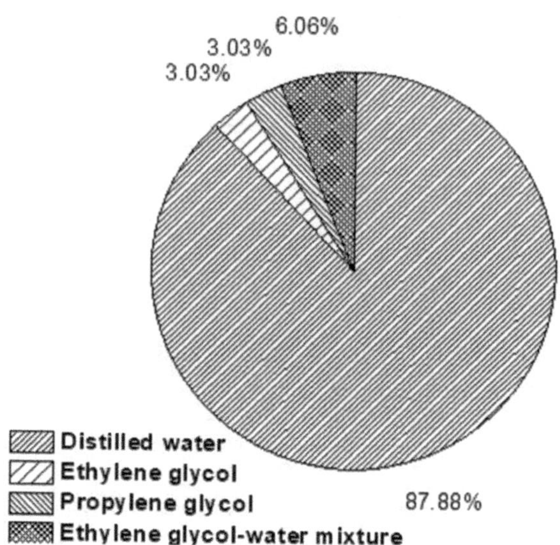

FIGURE 4.4 Utilization percentage of different base fluids for the preparation of nanofluids.

Due to its versatility and widespread use in various industries, the majority of research has focused on using distilled water as a base fluid for producing nanofluids. According to the data gathered from the Elsevier database spanning from 2010 to 2023, almost 87% of the published publications utilized water that had been distilled or deionized as the primary fluid for synthesizing nanofluids.[5] In addition, **Figure 4.4** illustrates the utilization of ethylene glycol–water mixture, ethylene glycol, and propylene glycol as alternative base fluids in diverse applications.

Choosing an appropriate base fluid for producing HNF is crucial for achieving a comprehensive enhancement of characteristics through various means.[6] The crucial thermophysical and optical parameters of any base fluid are thermal conductivity, viscosity, density, boiling point, freezing point, absorptivity, reflectivity, transmissivity, and extinction coefficient. For instance, ethylene glycol has higher viscosity compared to water and also possesses lesser thermal conductivity. Water is a more effective coolant than ethylene glycol for heat transfer applications. However, in colder regions, water cannot be utilized due to its freezing point of 0°C, while ethylene glycol freezes at −12°C. For an application involving the prevention of freezing, ethylene glycol is a superior choice of fluid compared to water.[7]

Note: Water and ethylene glycol, when mixed in a volume ratio of 30:70, will freeze at a temperature of −55°C.

The following list is a compilation of some crucial factors.

i. **Thermal conductivity**:
   The selection of the base fluid significantly affects the increase in thermal conductivity of the HNF. The thermal conductivity of base fluids varies depending on their composition, and this characteristic is enhanced by the inclusion of nanoparticles.

ii. **Viscosity**:
The viscosity of the base fluid is a crucial factor in influencing the flow characteristics of the nanofluid. The incorporation of nanoparticles can exert an influence on the viscosity of the HNF, and the nature of the base fluid will interact with these nanoparticles in a manner that impacts the overall viscosity.

iii. **Stability**:
The stability of nanofluid is of utmost importance for the practical utilization of HNFs. The choice of the base fluid can impact the stability of the nanofluid by modulating the interactions between the nanoparticles and the molecules of the fluid. Certain base fluids may exhibit superior stability, while others may lead to the aggregation of nanoparticles.[8]

iv. **Heat transfer enhancement**:
The performance can be influenced by the choice of base fluids, as different fluids may have variable effects on the capabilities of HNFs. Certain foundational fluids can enhance heat transfer by facilitating the effective dispersion and suspension of nanoparticles.

v. **Economic potential**:
The cost and availability of HNF production can also be influenced by the choice of base fluid. The selection of a specific base fluid can be influenced by issues such as cost, environmental concerns, and the accessibility of the fluid.

vi. **Point of application**:
The choice of the base fluid is contingent upon the particular demands of the application. Water-based nanofluids are frequently employed because of their eco-friendliness, although oil-based nanofluids may be favored in specific industrial applications.

vii. **Compatibility**:
Chemical compatibility is of utmost importance for the stability and effectiveness of HNFs, as it determines how well the base fluid and nanoparticles can coexist without any adverse reactions. The choice of the base fluid can impact the interactions between the fluid and nanoparticles, hence influencing the overall chemical stability.[9]

Example:
Carbon nanotubes and graphene nanoparticles, being hydrophobic, are incapable of dispersing in hydrophilic water. To enhance the dispersibility of nanoparticles in water, surfactants such as SDBS, CTAB, and sodium oleate are employed. Furthermore, the CNT nanoparticles undergo functionalization through the incorporation of a covalent bond into the carbon structure.

viii. **Pressure drop**:
The pressure drop properties of the HNF in a given system can be influenced by the type of base fluid used. Incorporating nanoparticles can modify the flow characteristics of the fluid, and different base fluids may demonstrate diverse pressure drop behaviors.[10]

TABLE 4.4
Type of Nanoparticles and Base Fluids from Well-Known Research Papers

Type of Nanoparticle	Base Fluid
Cu	Water, EG, oil, acetone, and water–EG mixture
Ag	Water and toluene
Au	Water, EG, and toluene
Al	Water, EG, oil, and kerosene
Zn	Vegetable oil, paraffin oil, and motor oil
Al_2O_3	Water, EG, oil, water, and glycerin mixture
CuO	Water, oil, transformer oil, and water–EG mixture
ZnO	Water, EG, oil, and water–EG mixture
TiO_2	Water, EG, oil, and water–EG mixture
SiO_2	Water, EG, glycerol, oil, and water–EG mixture
MWCNT	Water, EG, and water–EG mixture
DWCNT	
SWCNT	
Graphene	

The efficiency of an HNF is contingent upon various elements, such as the composition and concentration of nanoparticles, in addition to the unique demands of the application. Typically, researchers perform experimental studies and detailed analyses to optimize the formulation of HNFs according to the desired qualities for a certain application, and some of the examples are listed in **Table 4.4**.

4.6 TYPE AND CONCENTRATION OF SURFACTANT

Surfactants are crucial additives used to enhance the stability of colloidal suspensions for improving dispersion. The selection of surfactant type and concentration is essential and should be based on the types of base fluid and nanoparticles. Surfactants can be classified into several categories, including anionic, cationic, non-ionic, and amphoteric, as illustrated in **Figure 4.5**. Cationic and anionic surfactants are ionic dispersion agents characterized by polar molecules with opposite charges, with one end slightly positive and the other end slightly negative. When these charged molecules are present, the nanoparticles undergo significant electrostatic forces. Furthermore, the selection of the surfactant to be added depends on the charge of nanoparticles obtained in the base fluid. Typically, anionic surfactants are employed to disperse positively charged nanoparticles and vice versa. In contrast to ionic surfactants, non-ionic surfactants stabilize the dispersion using the steric hindrance effect without any electrical charges. The steric hindrance effect is created by adsorbing capping agents and polymer chains onto the surface of nanoparticles.

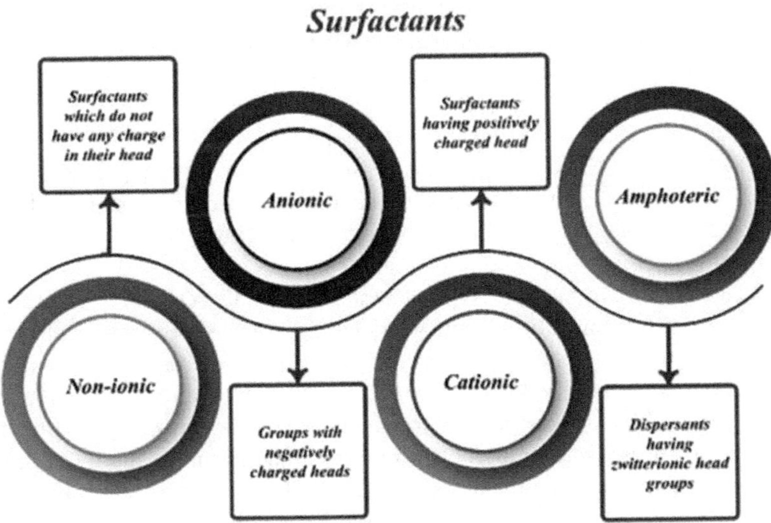

FIGURE 4.5 Types of surfactants which can be utilized for the stabilization of nanofluids.[11]

Moreover, non-ionic surfactants are used when the presence of ionic surfactants could disrupt the intended charge on the nanoparticles. Some of the non-ionic surfactants are as follows:

- Polyethylene glycol (PEG)
  Polyethylene glycol is a polymeric compound of ethylene glycol, and due to its hydrophilicity it's widely used in the dispersion of nanoparticles in water-based nanofluids. This polymeric chain is attached to the nanoparticle surface to minimize the van der Waals attraction between the particles.
- Polyvinyl alcohol (PVOH)
  PVOH is a hydrophilic synthetic polymer non-ionic surfactant that enhances the stability of the nanofluid by forming a barrier of protection around the nanoparticles. PVOH surfactants are commonly employed in water-based nanofluids due to their water solubility.
- Few more non-ionic surfactants
  Tween 20 and 80, Triton X-100, and TERGITOL NP-10 are the often-employed non-ionic surfactants to stabilize nanofluids. The steric stabilization mechanism is specifically employed for non-ionic surfactants. Repulsive forces are generated to promote the overlapping of nanoparticles.

4.7 SUMMARY

HNFs have been demonstrated to be preferable to nanofluids based on single nanoparticles because they exhibit a synergistic impact when nanoparticles are combined, resulting in more adjustable qualities. The enhancement of the properties is determined by various factors, such as the type and compatibility of the nanoparticles, a thorough comprehension of the ideal mixing ratio, the optimal concentration of the nanofluid, the selection of a suitable base fluid, and the stability mechanism involving the specific surfactant employed. Given the variations in features such as physical, mechanical, chemical, optical, electrical, shape, preparation procedures, and cost analysis among nanoparticles, the selection of the most suitable nanoparticle is of utmost importance. The authors of most studies selected nanoparticles based on their compatibility and suitability for the intended purpose. For instance, if the final application is a coolant for a heat exchanger, the nanoparticle must possess a high thermal conductivity.

After determining the specific type of nanoparticles, it is crucial to carefully evaluate the suitable mixing ratio of nanoparticles during the preparation of the HNF. However, many research have overlooked this aspect. One particle may possess a higher thermal conductivity but be more expensive than another particle that is relatively cheaper and has lower thermal conductivity. The optimization of this mixing ratio is crucial for getting the highest possible enhancement in the thermophysical properties while minimizing the associated cost. Therefore, it is advisable to utilize appropriate statistical tools to get optimization. Hence, the primary objective of optimizing the mixing ratio is to get the HNF at a reduced expense while enhancing its characteristics for convenient commercialization. In addition, the stability of HNFs, which consist of several nanoparticles, is of utmost relevance. Many researchers emphasize the significance of utilizing suitable surfactants at optimal concentrations to prevent agglomeration of the nanoparticles.

REFERENCES

1. Xuan, Z., Zhai, Y., Li, Y., Li, Z., & Wang, H. (2022). Guideline for selecting appropriate mixing ratio of hybrid nano fluids in thermal management systems. *Powder Technology, 403*, 117425. https://doi.org/10.1016/j.powtec.2022.117425
2. Mehrali, M., Sadeghinezhad, E., Latibari, S. T., Kazi, S. N., Mehrali, M., et al. (2014). Investigation of rheological and thermal conductivity properties of castor oil nanofluids containing graphene nanoplatelets. *Nanoscale Research Letters, 9*, 1–12.
3. Moghaddam, M. H., & Karami, M. (2022). Heat transfer and pressure drop through mono and hybrid nanofluid-based photovoltaic-thermal systems. *Energy Science & Engineering, 10*, 918–931.
4. Malika, M., & Sonawane, S. S. (2021). Low-frequency ultrasound assisted synthesis of an aqueous aluminium hydroxide decorated graphitic carbon nitride nanowires based hybrid nanofluid for the photocatalytic H2 production from Methylene blue dye. *Sustainable Energy Technologies and Assessments, 44*, 100979.
5. Malika, M., & Sonawane, S. S. (2021). Statistical modelling for the Ultrasonic photodegradation of Rhodamine B dye using aqueous based Bi-metal doped TiO2 supported montmorillonite hybrid nanofluid via RSM. *Sustainable Energy Technologies and Assessments, 44*, 100980.
6. Malika, M., & Sonawane, S. S. (2021). Application of RSM and ANN for the prediction and optimization of thermal conductivity ratio of water based Fe2O3 coated SiC hybrid nanofluid. *International Communications in Heat and Mass Transfer, 126*, 105354.
7. Malika, M., & Sonawane, S. S. (2021). A comprehensive review on the effect of various ultrasonication parameters on the stability of nanofluid. *Journal of Indian Association for Environmental Management (JIAEM), 41*(4), 19–25.

8. Sonawane, S. S., & Malika, M. (2021). Review on CNT based hybrid nanofluids performance in the nano lubricant application. *Journal of Indian Association for Environmental Management (JIAEM)*, *41*(3), 1–16.
9. Malika, M., & Sonawane, S. S. (2021). The sono-photocatalytic performance of a novel water based Ti+ 4 coated Al (OH) 3-MWCNT's hybrid nanofluid for dye fragmentation. *International Journal of Chemical Reactor Engineering*, *19*(9), 901–912.
10. Malika, M., Bhad, R., & Sonawane, S. S. (2021). ANSYS simulation study of a low volume fraction CuO–ZnO/water hybrid nanofluid in a shell and tube heat exchanger. *Journal of the Indian Chemical Society*, *98*(11), 100200.
11. Yang, L., Ji, W., Mao, M., & Huang, J. (2020). An updated review on the properties, fabrication and application of hybrid-nanofluids along with their environmental effects. *Journal of Cleaner Production*, *2020*, 120408. https://doi.org/10.1016/j.jclepro.2020.120408

5 Hydrothermal and Rheological Properties of Hybrid Nanofluids

Hydrothermal characteristics of nanofluids are the behavior of nanofluids under conditions of temperature and pressure that have been described. In addition to thermal conductivity, thermal stability, heat transfer enhancement, boiling, and condensation, these properties include thermal conductivity. It is absolutely necessary to have a strong understanding of these properties to be able to provide reproducible augmented properties for real-time industrial heat transfer applications. In the following, the rheological and hydrothermal properties, as well as some of the most important variables are discussed.

5.1 HYDROTHERMAL PROPERTIES OF NANOFLUIDS

5.1.1 Thermal Conductivity

Based on an examination of thermal properties, it is clear that all fluids demonstrate exceptionally low thermal conductivity. However, it should be noted that liquid metals are an exception to this rule, as they are not suitable for use within most practical temperature ranges. Water's thermal conductivity is around 60 orders of magnitude lower than that of aluminum oxide (Al_2O_3) when it comes to heat conduction. The same applies to radiator coolants and lubricants as well. Undoubtedly, the inherent limitation of the fluid's thermal conductivity will inevitably impede any attempts to improve heat transmission through means such as expanding surface area, producing turbulence, or employing similar techniques. Therefore, it is logical to expect that efforts will be made to enhance the thermal conduction properties of coolants. Over a century ago, one potential solution that was considered was the utilization of solid suspension. In 1873, J.C. Maxwell[1] presented a theoretical framework and a set of mathematical equations, specifically **Equation 5.1**, for determining the effective thermal conductivity of a suspension. A constraint of this model is its applicability exclusively to particles with a spherical shape.

$$\frac{\frac{K}{K_1}-1}{\frac{K}{K_1}+2}=V_2\left(\frac{\frac{K_2}{K_1}-1}{\frac{K_2}{K_1}+2}\right) \tag{5.1}$$

Here,

K: Conductivity of the mixture

K_1: Conductivity of the continuous phase (base fluid)

K_2: Conductivity of the discontinuous phase (particles)

V_2: Volume fraction of the particles

DOI: 10.1201/9781003595137-5

In continuation of Maxwell's theory, H. Fricke[2] expanded upon his research and formulated a mathematical equation, known as **Equation 5.2**, to describe the effective conductivity of ellipsoid-shaped particles in a heterogeneous system in the year 1924. The established equation was employed to assess the conductivity of experimental data from dog blood, and it accurately detected nearly 90% of the data.

$$\frac{\frac{K}{K_1}-1}{\frac{K}{K_1}+X}=V_2\left(\frac{\frac{K_2}{K_1}-1}{\frac{K_2}{K_1}+X}\right) \tag{5.2}$$

Here, X is the shape factor of a particle, which is strongly a function of K_1 and K_2 unless there is a large difference in the conductivity of continuous and discontinuous phases.

If $X = 2$, **Equation 5.2** is reduced to **Equation 5.1**.

Subsequent to their work, a plethora of theoretical and experimental research was undertaken, including studies conducted by Hamilton–Crosser and Wasp.

5.1.1.1 Hamilton–Crosser Theory

Hamilton and Wasp[3] developed model equations to determine the thermal conductivity of slurries, which are two-phase systems in 1962. More precisely, the theory under consideration establishes a correlation between the efficiency of thermal energy transfer in diverse systems, such as the conduction of heat through ceramics, glasses, and packed beds. Furthermore, in the context of heat transfer performance in heterogeneous systems including solid bodies, conduction assumes a more prominent role compared to convection and radiation. The experiment involved dispersing a non-continuous phase, composed of a silicone elastomer, into a continuous phase, composed of solid aluminum particles with various unique forms. Furthermore, the results revealed the importance of particle morphology in enhancing thermal conductivity. For example, the thermal conductivity of cylindrical aluminum particles is roughly twice as high as that of spherical materials. **Equation 5.3** contains a depiction of the Hamilton and Wasp model equation. This equation can be applied to any possible form of particles. The reproducibility of the data was discovered to be approximately 2%. It is vital to mention that each of these tests focused on suspending particles of various sizes, ranging from micron to micro size.

$$K=K_1\left[\frac{K_2+((n-1)K_1)-((n-1)V_2(K_1-K_2))}{K_2+((n-1)K_1)-(V_2(K_1-K_2))}\right] \tag{5.3}$$

Here,

$$n=\frac{3}{\text{Sphericity of the particle}}$$

Note: n is not a function of the shape of the particle unless there is a difference between the conductivities of continuous and discontinuous phases.

Limitations:

- The introduction of a discontinuous phase of particles into a continuous phase of base fluid would undoubtedly result in the formation of an interface between the two phases. It is therefore possible that the shape of the particles will be altered as a result of this interaction; nevertheless, there are no definitive studies that can be submitted to refute this assertion.
- In addition, the experimentation presents a significant challenge in terms of practically maintaining monochromatic particles with the same shape.
- Because of this, the system that was designed might not produce reproducible outcomes.

5.1.1.2 Wasp Theory

In their book, Wasp et al.[4] have provided practical guidelines for the design of pumps, valves, and pipelines to ensure the effective transportation of slurries. Here is the synopsis of the book. Based on the survey, slurry transportation commenced in 1970, greatly facilitating the transportation of minerals for the mineral industries. To ensure efficient transportation, it is necessary to comprehend the hydraulic characteristics of slurry. The distinct characteristics of particles and the underlying fluid have a significant impact on the properties of the resulting combination. Furthermore, the operating conditions significantly influence the behavior of slurry. To successfully industrialize nanofluids, it is important to analyze the transportation issues of slurries, as nanofluids are essentially colloidal suspensions of nanoparticles in a base fluid. By doing so, appropriate design conditions may be determined and documented.

For the proper designing of slurry transportation, it is necessary to study its physical properties as a function of individual fluid and the slurry as a continuous phase. In this chapter, we discuss the most important properties of the suspension and its measurement techniques. Nanofluids consist of nanoparticles suspended in a base fluid medium. The density of a nanofluid is determined by three factors: the density of the nanoparticles, the density of the base fluid, and the overall density of the nanofluid. Typically, the particle's density is already provided. In the absence of data, the density can be manually measured using the basic specific gravity bottle. If the particle size exhibits heterogeneity, it is advisable to take into account the mean value. A specific gravity bottle can be used to measure the density of nanofluids, as long as the suspension is homogenous and stable, without any formation or settling of aggregates. Currently, there is a wide range of slurry density meters available in the market that can accurately measure the density of slurries in real time. This device utilizes a non-invasive ultrasonic sensor to accurately measure the instantaneous density of mineral slurries in high-density conditions and within large pipes.

Theoretically, the density of the suspension can be measured using the following equation:

$$\rho_{\text{Slurry}} = \frac{100}{\dfrac{C_{\text{Solids}}}{\text{Density of solid particle}} + \dfrac{100 - C_{\text{Solids}}}{\text{Density of fluid}}} \tag{5.4}$$

Here, C_{Solids}: Concentration of the solid particles (Weight %)

$$\text{Also, Concentration of the solid particles } C_V\left(Volume\ \%\right) = \frac{C_{\text{Solids}}\,\rho_{\text{Solid}}}{\rho_{\text{Fluid}}} \tag{5.5}$$

$$\text{Solid volume fraction } \left(\varphi - Volume\ \%\right) = \frac{C_V}{100} \tag{5.6}$$

5.1.1.3 Importance of Using Nanofluids with High Thermal Conductivity in Heat Transfer Applications

When the nanoparticles are dispersed in the base fluid, it is anticipated that the resulting fluid has improved properties due to the synergistic effect between the particles and base fluid. With these improvements, there are numerous advantages in the following applications.

5.1.1.3.1 High Heat Conduction

The addition of nanoparticles with a high surface-to-volume ratio has increased the overall heat transfer area, which enables them to transfer the heat effectively with the utilization of thermal bridging phenomena as shown in **Figure 1.2**. In general case, while preparing the hybrid nanofluid, the average particle size of nanoparticles should be less than 100 nm and less than 20 nm is preferable. Lower the particle size lower the space between the molecule's interaction more effective the transfer of heat. The random collisions between the particles initiate the continuous Brownian motion of the particles to achieve the synergistic effect. Due to the addition of nanoparticles, enhanced random motion induced the accelerated micro-convection in the fluid to increase the rate of heat transfer. Further increase in temperature has increased the kinetic energy of the particles to enhance thermal conductivity around three folds than at room temperature. This increment was seen only up to some limit of temperature increment due to the aggregation of the particles at elevated temperatures. So, it is evident that maintaining the stability of nanofluid is one of the significant limitations of industrialization.

This mobility has the potential to induce micro-convection of fluid, which therefore increases heat transfer. There is a possibility that the micro-convection and greater heat transfer will further accelerate the rate at which heat is dispersed throughout the fluid. It has already been discovered that the thermal conductivity of nanofluids dramatically rises with an increase in temperature. This phenomenon, which may be related to the causes mentioned earlier, has been observed. It is stable. The fact that the particles are so small means that they weigh less, and this also means that the likelihood of sedimentation is reduced. By reducing sedimentation, it is possible to circumvent one of the most significant limitations of suspensions, which is the settling of particles, and to make nanofluids more consistently stable.

Microchannel cooling does not allow clogging. Not only will nanofluids be an improved medium for heat transmission in general, but they will also be an excellent choice for microchannel applications that include the presence of significant heat loads. The utilization of nanofluids in conjunction with microchannels will result in the production of highly conductive fluids and a substantial area for heat transfer. This is not possible with meso- or microparticles because they clog microchannels, which prevents them from occurring. Nanoparticles, which consist of only a few hundreds or thousands of atoms, are noticeably smaller than microchannels, which are several orders of magnitude larger.

While nanoparticles are extremely small, the amount of velocity that they are able to transfer to a solid wall is significantly less. Consequently, the likelihood of components like heat exchangers, pipes, and pumps being eroded is decreased as a result of this decreased velocity. Pumping power typically needs to be raised by a factor of 10 to get a twofold increase in the amount of heat that can be transferred from traditional fluids. If one is able to multiply the conductivity by a factor of three, it is possible to demonstrate that the heat transfer in the same device will increase by a factor of two. Unless there is a significant rise in the viscosity of the fluid, the required increase in pumping power will be quite small. Therefore, if a significant increase in thermal conductivity can be obtained with a very small volume fraction of particles, then it is possible to accomplish a very big reduction in the amount of power that is required for pumping.

5.2 RHEOLOGICAL PROPERTIES OF NANOFLUIDS

The majority of real fluids are viscous fluids that exhibit resistance to motion due to internal shearing within the fluid. This phenomenon is studied in the field of rheology. Viscosity is primarily caused by intermolecular friction among fluid molecules, which must be overcome to commence fluid movement. **Figure 5.1** illustrates the rheological characteristics of fluids. **Equation 5.7** expresses Newton's law of viscosity, which establishes a direct correlation between shear stress and shear strain. Fluids that adhere to Newton's law of viscosity are referred to as Newtonian fluids. In a Newtonian fluid, the relationship between shear stress and shear rate, shear strain, or velocity gradient is linear, and the constant of proportionality is known as viscosity (μ; **Equation 5.8**). Non-Newtonian fluids are those whose viscosity varies with temperature, time, and shear rate, rather than remaining constant. These fluids are highly viscous, and

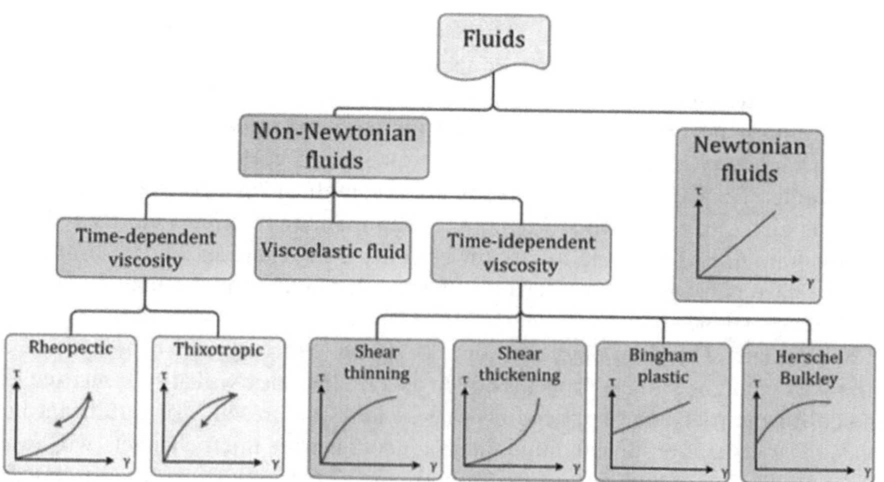

FIGURE 5.1 Shear stress versus shear rate or velocity gradient for Newtonian and non-Newtonian fluids.[5]

their viscoelasticity is very important. Non-Newtonian fluids can be classified as either time-dependent or time-independent. Shear thickening fluids, also known as dilatant fluids, refer to fluids that experience an increase in viscosity as the shear rate increases. On the contrary, shear-thinning fluids, or pseudoplastic fluids, are fluids that exhibit a drop in viscosity as the shear rate increases. Corn starch in water and blood are prime illustrations of dilatant and pseudoplastic fluids, respectively.

$$\text{Shear stress}(\tau) \propto \text{Velocity gradient}\left(\frac{du}{dy}\right) \tag{5.7}$$

$$\tau = \mu\left(\frac{du}{dy}\right) \tag{5.8}$$

Here,
τ: Shear stress (N/m^2)
μ: Absolute viscosity (Pa. sec.)
$\frac{du}{dy}$: Velocity gradient in a straight or a parallel flow (sec^{-1}). Velocity gradient is also called as shear strain or shear rate (γ).

The inclusion of nanoparticles increased the viscosity of the base fluid, despite its superior properties. At high concentrations of nanofluids, the intermolecular interactions between the nanoparticles are enhanced, which can lead to a more pronounced effect in this scenario. However, in an industrial setting, it is crucial to have a well-defined understanding of the flow characteristics of the nanofluid, specifically in terms of viscosity. The higher the viscosity, the greater the resistance to flow, resulting in increased pumping power needed to transfer the fluid. This leads to additional costs for the process. The increase in viscosity can depend on several parameters, such as the type of base fluid, the type of nanoparticle, its morphology, concentration, and cost. It can also be influenced by operating conditions, including temperature, pH of the solution, and the type and concentration of the surfactant. If the generated nanofluid has poor stability, an increase in particle aggregation can be observed at high nanofluid concentrations, which directly impacts the nanofluid's performance and increases pumping costs.

There are numerous mathematical equations that can be utilized to forecast the flow characteristics of nanofluids in terms of viscosity. Given that nanofluids consist of nanoparticles in low concentrations, it is crucial to analyze the rheology of the nanofluid by treating it as a dilute solution. Around the year 1905, Einstein formulated an equation to quantify the absolute viscosity of a dilute suspension (**Equation 5.9**).

$$\frac{\mu_{\text{Slurry}}}{\mu_{\text{Base fluid}}} = 1 + 2.5\varphi \tag{5.9}$$

Einstein model is applicable for

- Laminar flow
- Spherical-shape particles

- Dilute suspension with solid volume fraction less than 0.02 Vol%
- No interaction between the particles are expected due to the dilute nature of the suspension.

Following Einstein's viscosity equation, Mooney[6] produced a model equation in 1951 to estimate the viscosity of spherical particles. This equation takes into account the interactions between the particles, which are influenced by the particle crowding factor. **Equation 5.10** can be applied to a broad range of nanofluid concentrations and distributions of poly-dispersed particles.

$$\frac{\mu_{\text{Slurry}}}{\mu_{\text{Base fluid}}} = e^{\left(\frac{1+2.5\varphi}{1-k\varphi}\right)} \tag{5.10}$$

Here, k represents the self-crowding factor, which may be determined based on experimental data by considering the crowding of spheres with different particle radii, r_{i} and r_{j}.

Brinkman[7] has presented a straightforward method in an independent investigation to forecast the viscosity of suspensions, even when there are significant levels of solid particles, based on the known values at infinite dilutions (**Equation 5.11**). If there are a total of "n" particles present in a base fluid with a volume of "V", the suspension's concentration is determined by $\left(\frac{n}{v}\right)$. If an additional particle of volume (v_0) is introduced to the suspension, the viscosity will increase as a result of crowding.

$$\frac{\mu_{\text{Nanofluid}}}{\mu_{\text{Base fluid}}} = \frac{1}{\left(1-C_v\right)^{2.5}} \tag{5.11}$$

Here, $C_V = \frac{nv_0}{V}$

Thomas[8] has devised a model equation (**Equation 5.12**) to overcome the constraints identified by Einstein in 1965. This equation accurately measures the viscosity of a concentrated dispersion, taking into account the interactions between particles. The equation yields a variance of 0.152.

$$\frac{\mu_{\text{Slurry}}}{\mu_{\text{Base fluid}}} = 1+2.5\varphi+10.05\varphi^2+0.00273e^{16.6\varphi} \tag{5.12}$$

The aforementioned equation was formulated to tackle the difficulties associated with the assumption that a concentrated suspension, rather than a dilute solution, is being studied.

- In the experimental setup, the hydrodynamic contact between the particles is also taken into account.
- The interactions are primarily unaffected by parameters such as flocculation, particle size, particle size distribution, and the type of viscometer.

Lundman[9] has developed a correlation in an independent investigation to predict the effective viscosity of flow through a stationary bed and a suspension of any concentration. The particles have a set, solid spherical shape, but the statistical analysis allows for the study to be expanded including particles of any shape. **Equation 5.13** provides the correlation for the effective viscosity of nanofluid.

$$\frac{\mu_{\text{Nanofluid}}}{\mu_{\text{Base fluid}}} = \frac{1}{1-2.5\varphi} \tag{5.13}$$

Researchers have only examined the impact of solid volume fraction on the effective viscosity of a solution so far. In 1977, Batchelor[10] proposed the impact of Brownian motion of a stable nanofluid, in addition to solid volume fraction. Here, the author has assumed that the nanoparticles are spherical. **Equation 5.14** correlation has been derived from numerical testing of weak and strong Brownian motion in a stable suspension considering all the nanoparticles are in the same direction. In the scenario of weak Brownian motion of particles in dilute suspensions, the value of φ^2 is estimated to be 7.6.

$$\frac{\mu_{\text{Nanofluid}}}{\mu_{\text{Base fluid}}} = 1+2.5\varphi+6.2\varphi^2 \tag{5.14}$$

Graham[11] has examined how nanoparticle diameter and spacing between spherical particles impact the relative viscosity of nanofluid after analyzing Brownian motion. The derivation includes energy dissipation from particle interactions and fluid movement around spherical particles. **Equation 5.15** provides the estimated model equation. Here, h represents the distance between the particles and D_{P} represents the diameter of the particles.

$$\frac{\mu_{\text{Nanofluid}}}{\mu_{\text{Base fluid}}} = \mu_{\text{Relative}} = 1+2.5\varphi+\frac{4.5}{\left[\frac{h}{D_{\text{P}}}\left(2+\frac{h}{D_{\text{P}}}\right)\left(2+\frac{h}{D_{\text{P}}}\right)^2\right]} \tag{5.15}$$

Despite the existing model equations, there are multiple inconsistencies between the experimental results and the extrapolated data. Not all the researchers will utilize identical nanoparticles with uniform size, shape, and purity. Agglomeration of nanoparticles onto one other can occur at high concentrations of nanofluids. Aside from nanofluid concentration, other aspects such as the base fluid type, operation temperature, nanoparticle type and shape, solution pH, and stability mechanisms can also influence the relative viscosity of nanofluid. Masoumi et al.[12] developed a model equation in the year of 2009 for the relative viscosity of a nanofluid, taking into account variables such as Brownian motion of particles, type, size, and properties of nanoparticles (such as density), concentration of nanofluid parameters, and distance between particles. The experimental results for alumina nanoparticles measuring 13 nm and 28 nm were compared with the model equation.

The study calculated Brownian motion (V_{B}), Reynolds number due to Brownian velocity (Re_{B}) for the nanoparticles in Brownian motion, and the distance between the particles using the following equations. **Equation 5.19** provided the viscosity of the nanofluid.

$$\text{Brownian velocity}\left(V_B\right) = \frac{1}{D_P}\sqrt{\frac{18K_BT}{\pi\rho_P D_P}} \tag{5.16}$$

$$\text{Renolds number due to Brownian velocity} = Re_B = \left(\frac{\rho}{\mu}\right)_{bf} V_B D_P \tag{5.17}$$

$$\text{Distance between particles} = h = \sqrt[3]{\frac{\pi D_P}{6\varphi}} \tag{5.18}$$

$$\mu_{nf} = \mu_{bf} + \frac{\rho_P V_B {D_P}^2}{72Ch} \tag{5.19}$$

$$\text{Correlation factor}\left(C\right) = \frac{1}{\mu_{bf}}\begin{bmatrix}\varphi\left(-1.133\times10^{-6}D_P - 2.771\times10^{-6}\right)\\ +\left(D_P\times9\times10^{-8} - 3.93\times10^{-7}\right)\end{bmatrix} \tag{5.20}$$

In the expression,

K_B: Boltzmann's constant
T: Temperature of nanofluid
ρ_P: Density of nanoparticle
D_P: Average diameter of nanoparticles
V_B: Brownian velocity
φ: Solid volume fraction of nanofluid

Subscript:

bf: Base fluid
nf: Nanofluid

Upon testing the developed model equation with the CuO/water nanofluid experimental data and other existing models, a high level of agreement was discovered, as depicted in **Figure 5.2**.

5.2.1 Effect of Viscosity of Nanofluid on the Pressure Drop and Pumping Power

5.2.1.1 Case 1: Fluid Flow in Heat Exchangers

Pressure drop occurs when there is a decrease in pressure while a fluid flows through a system. It is crucial to consider this element while designing systems for real-time industrial applications as it directly impacts the fluid flow rate, pumping power, and total heat transfer efficiency of a heat exchanger. Several factors affect the pressure drop in a heat exchanger in real time. The main elements include the characteristics of the fluid itself and the heat exchanger's geometry.

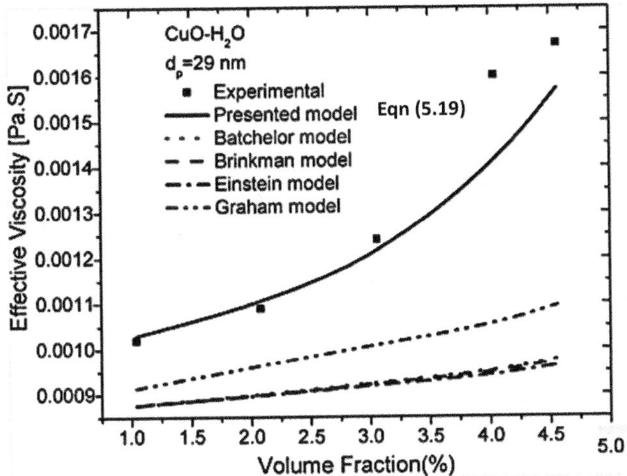

FIGURE 5.2 Comparison of experimental results of CuO/water nanofluid with available model equations for the prediction of viscosity of nanofluid.

Pressure drop is induced by several factors, including:

- Fluid flowing through channels or heat exchangers experiences frictional resistance against the walls due to viscosity. This resistance to flow results in a decrease in pressure, known as pressure drop.
- Fluid characteristics such as viscosity and density are crucial elements that contribute to increased resistance to flow, thereby impacting the pumping power and pressure drop in fluid flow. An increase in fluid frictional loss is a crucial aspect to take into account.
- Additionally, the hot and cold fluids travel via channels, experiencing shifts and turns that result in velocity dispersion. Fluid flowing in various directions may undergo expansions and contractions, leading to pressure drops. Moreover, these alterations in velocity impact the pressure drop.
- Fluid flow through obstacles such as valves and fittings also leads to an increase in pressure loss.
- The intricate shapes of process systems such as shell and tube and plate-type heat exchangers lead to an increase in pressure loss due to the flow around the design.
- Transitioning from laminar flow to turbulent flow results in increased pressure loss in the flow due to turbulence and frictional forces.

5.2.1.2 Case 2: Nanofluid Flow in Heat Exchangers

When nanoparticles are disseminated in a colloidal solution together with conventional fluids, they affect the frictional losses in fluid flow due to changes in fluid characteristics. The particle's interaction has impacted both the fluid's flow behavior

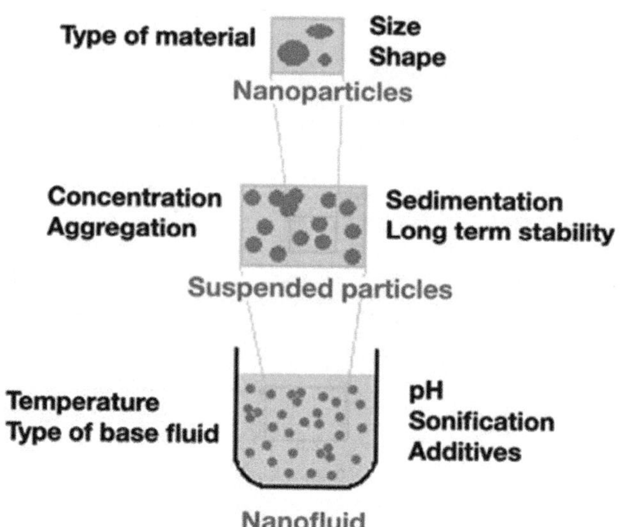

FIGURE 5.3 Effect of nanoparticle type, size, and shape on the stability of nanofluid colloidal suspension.[13]

and its properties. The interactions of nanoparticles are affected by their size, shape, and surface chemistry as shown in **Figure 5.3**. Some of the important factors are given in the following subsections.

5.2.1.2.1 *Effect of Size of the Particle*

Small particles have a high surface-to-volume ratio, which leads to reduced agglomeration and settling. Nanofluids and hybrid nanofluids require at least one dimension to be less than 100 nm, preferably under 20 nm for practical use. Prior to industrializing the specific nanofluid based on nanoparticle size, an optimization analysis is necessary. The smaller the size, the higher the product cost. Additionally, nanoparticles must be in a colloidal suspension condition when added to the base fluid. Viscosity can increase even at low concentrations if the particles are bigger in size.

5.2.1.2.2 *Effect of Surface Properties of Nanoparticles*

The surface properties of nanoparticles consist of anionic and cationic characteristics. Typically, nanoparticles adsorb ions from the base fluid onto their surface and acquire a surface charge based on the solution's pH. The stable dispersion mechanism of nanoparticles is determined by the hydrophilic and hydrophobic properties of the base fluid, and the inclusion of suitable surfactants is recommended.

5.2.1.2.3 *Effect of Shape of the Nanoparticle*

The shape of particles directly affects the density and inter-particulate interaction among nanoparticles in the colloidal solution. Rigorous investigations have demonstrated that nanoparticles of a spherical form exhibit higher stability over a prolonged period due to their limited packing capability and reduced agglomeration rate compared to particles of other shapes. However, further research is required to substantiate this argument.

5.2.2 Effect of Nanofluid Type and its Concentration on the Pumping Power of the System

The relationship between pressure drop, pumping power, and improved heat transfer rate depends on the Reynolds number (N_{Re}) of the turbulent flow. At larger concentrations of nanofluid, there is a risk of particles clumping together and settling, which should be prevented to ensure consistent results when measuring thermophysical parameters. It is essential to uphold the precise stability of nanofluid to ensure the preservation of transport capabilities. Optimizing nanofluid concentration is crucial for improving thermophysical characteristics. For instance, the thermal conductivity of a nanofluid rises with an increase in the concentration of nanoparticles in the base fluid. However, the increased amount of nanoparticles being added is significantly increasing the overall cost of the process, highlighting the importance of gaining a thorough grasp of this issue. For instance, there is a notable increase in the viscosity of nanofluid, leading to higher pressure drop and pumping power. Higher agglomeration rates of nanoparticles necessitate the use of optimization methods, even at an industrial scale.

High viscosities can lead to increased resistance to fluid flow, potentially causing blockages in microchannels and accelerating fouling on heat transfer surfaces. Using nanofluids may still reduce the heat transfer efficiency of the heat exchanger. When using nanofluids containing specific nanoparticles, particularly metals and metal oxides, the inner surface of the channel may corrode or erode due to the oxidation of metal nanoparticles at high temperatures over time. To address the pumping power issue, greater capacity pumps are necessary, even for smaller applications, which could lead to a higher overall system cost. Special attention should be given while selecting nanoparticles for use in producing nanofluids, considering their properties, size, and cost implications.

The pumping power for turbulent flow in a pipe of diameter "*D*" and velocity "*V*" is often calculated as follows:

$$\text{Pumping power}\left(\dot{P}\right)=\frac{\pi}{4}D^2V\left(\Delta P\right) \tag{5.21}$$

where

$$\text{Pressure drop}\left(\Delta P\right)=\frac{fL\rho V^2}{2D}$$

Here, f = friction factor $=\dfrac{0.3164}{N_{Re}^{\ 0.25}}$; $3000 < N_{Re} < 10^5$

L = Length of the pipe
ρ = Density of fluid
V = Velocity of the fluid
D = Diameter of the pipe
N_{Re} = Reynolds number

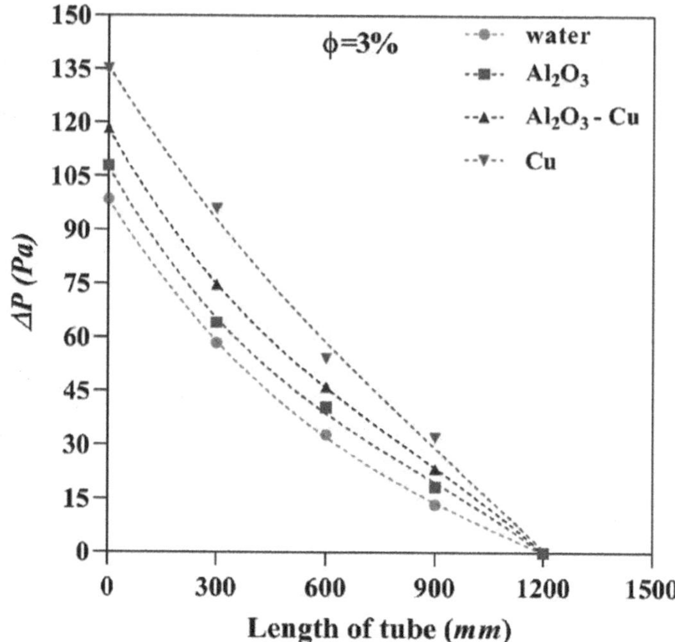

FIGURE 5.4 Effect of nanofluid type on the pressure drop of a double pipe heat exchanger.[14]

El-Magid Mohamed et al.[14] conducted a study on how nanofluid and hybrid nanofluid impact the efficiency of a twin pipe heat exchanger. The study evaluated the impact of nanofluid type (water-based Cu, Al_2O_3, and Cu-Al_2O_3), concentration, flow rate, and inner-tube rotation. Increasing the nanofluid concentration from 1 to 3 Vol% improved the heat transmission capability, as indicated by the results. Hybrid nanofluid does not always outperform mono nanofluid. In this instance, Al_2O_3/water exhibited superior heat transfer effectiveness compared to other nanofluids. Additionally, as the nanofluid concentration increased from 1 to 3 Vol%, the viscosity of the fluid, fluid friction, and pressure drop increased (**Figure 5.4**). The pumping power increased with the rise in pressure drop and fluid density, as depicted in **Figure 5.5**. Furthermore, the pumping power remained consistent at approximately 0.1 W for all nanofluid types, even at a concentration of 3 Vol%. This level of power does not strain the pump connected to the heat exchanger.

5.2.2.1 Important Point to Be Noted

To double the heat transfer performance of a coolant utilizing conventional fluids such as water, the pumping power typically needs to be raised by a factor of 10. Increasing thermal conductivity by a factor of three will double the heat transmission within the device. If the viscosity of the fluid remains relatively constant, the increase in pumping power needed will be minimal. A small percentage of particles, less than 1 Vol% (nanofluid concentration), can significantly enhance thermal conductivity, resulting in a considerable decrease in required pumping power.

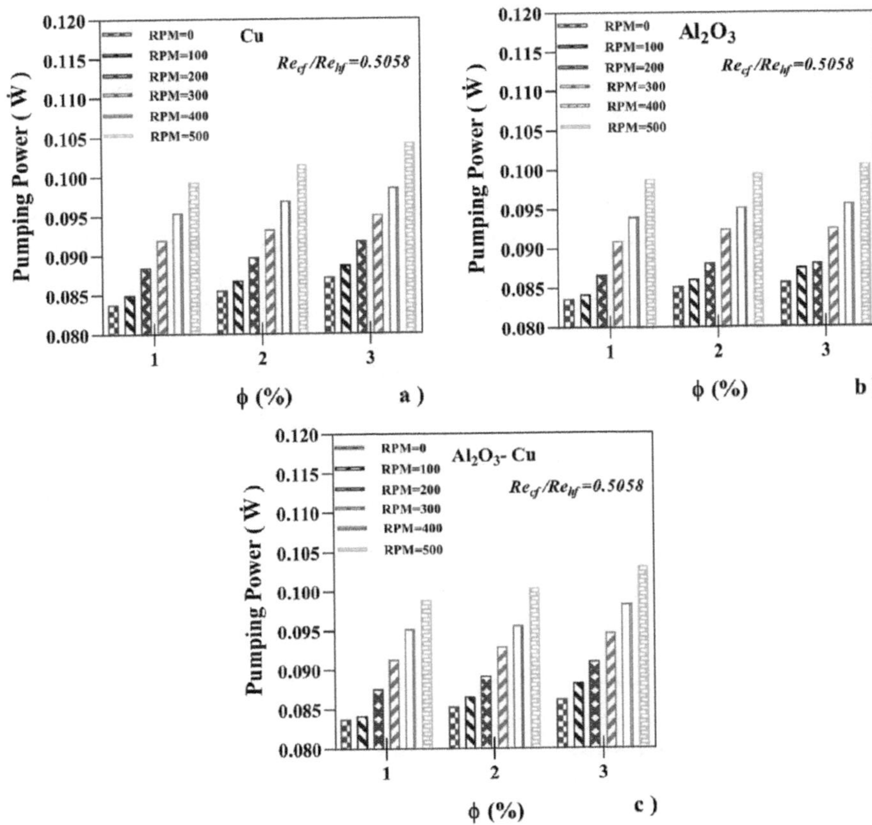

FIGURE 5.5 Effect of nanofluid type and its concentration on the pumping power of the system.[14]

5.2.2.2 Future Work

Prior research has mostly concentrated on increasing the heat transfer rate with minimal consideration for pressure drop. Hence, it is recommended to evaluate and extensively examine the effect of combined passive approaches on pumping power in conjunction with the augmentation of heat transfer rate.

5.3 HYDRAULIC CHARACTERISTICS OF SLURRY (COLLOIDAL SUSPENSION)

Fluids are classified into homogeneous and heterogeneous flow systems based on the arrangement of particles in the base fluid. In the homogeneous flow, particles are evenly distributed throughout the liquid media. The transition velocity is the point at which turbulent flow in fluid transportation operations changes to laminar flow. To maintain a consistent discharge pressure, it is advisable to operate at a velocity below the transition velocity while transporting homogenous fluids, as this reduces the risk of accidents. Furthermore, turbulent flow is the optimal choice for transporting fluids across long distances, even in systems with varying compositions. Several parameters must be taken into account while developing a slurry transportation system. The following are some of the process considerations for the designing.

i. **Hydraulics:**
   - Identifying the suitable base fluid for transportation. In regions with low temperatures, water is not suitable for usage due to its freezing point. Instead, antifreeze coolants such as ethylene glycol, propylene glycol, and a combination of water and ethylene glycol are utilized.
   - The optimization of particle size and concentration is crucial for selecting the most suitable option. Flow patterns, whether homogeneous or heterogeneous, are generated depending on the particle size. It is crucial to ensure the particle size remains homogeneous. Moreover, when the particle concentration is elevated, it might cause an increase in the fluid's viscosity in the vertical direction of a pipe. Therefore, optimizing the particle concentration is a crucial element.
   - After establishing the aforementioned parameters, it is necessary to conduct pilot-scale trial runs to ascertain the optimal operating velocity of the fluid.
   - Investigate the impact of pumping parameters on the properties of particles, such as their size and deterioration.

ii. **Surface failure testing:**
   - The interaction between minerals, particularly mineral oxides, leads to surface degradation such as corrosion, erosion, and wear in pipelines, which can be used to predict the lifespan of the pipeline.
   - Thorough studies on the abrasion of valves and pumps are essential.

iii. **Pump selection criteria:**
   The selection of pumps is determined by various factors such as design, location, space availability, maintenance requirements, reliability, and hydraulic needs.
   - Pump capacity (in cubic meters per hour)
   - Discharge head (in meters) that is necessary
   - Viscosity of the fluid (measured in centipoise)
   - Density of the fluid (in kilograms per cubic meter)
   - The weight percentage of solid particles present in the fluid being pumped
   - Fluid's vapor pressure (kPa)
   - Net positive suction head (NPSH) available, measured in meters
   - Operating temperature (°C)
   - Quantity and positions of pumps

iv. **Economic factors:**
   - The investment and operational costs
   - Additionally, costs associated with maintenance and the use of inhibitors
   - Expense for pilot plant studies

5.4 SUMMARY

This chapter explored the hydrothermal and rheological properties of hybrid nanofluids. Hydrothermal properties include the effect of elevated temperature and pressure on the thermal conductivity of nanofluids, whereas rheological properties include the study of fluid flow properties like viscosity and friction factor. Further, the effect of nanofluid type, particle size, and nanofluid concentration on the hydrothermal and rheological properties of hybrid nanofluids have been explored. Studies have shown that increases in nanoparticle size have shown the increment in the agglomeration of the nanoparticles and reduction of stability of nanofluid. Also, increases in nanofluid concentration have shown an increment in the thermal conductivity of the nanofluid, but at concentrations higher than 1 Vol%, the viscosity was also affected. With the increased viscosity, pressure drop and pumping power of the system can be elevated. So, maintaining an optimum nanofluid concentration of less than 1 Vol% is preferable.

Moreover, systematic approaches, standards, and guidelines are required so that consistent performance can be achieved. Collaborative efforts between academia and industries will play a crucial role in the commercialization of the nanofluids.

REFERENCES

1. Maxwell, J. C. (1873). *A Treatise on Electricity and Magnetism* (Vol. 1). Clarendon Press.
2. Fricke, H. (1924). A mathematical treatment of the electric conductivity and capacity of disperse systems I. The electric conductivity of a suspension of homogeneous spheroids. *Physical Review Journals Archive*, *24*, 575–587.
3. Hamilton, R. L. (1962). Thermal conductivity of heterogeneous two-component systems. *Industrial & Engineering Chemistry Fundamentals*, *1*, 187–191.
4. Wasp, E. J., Kenny, J. P., & Gandhi, R. L. (1977). *Solid-Liquid Flow Slurry Pipeline Transportation [Pumps, Valves, Mechanical Equipment, Economics]*. Technical Publications.
5. Khodadadi, H., Aghakhani, S., Majd, H., Kalbasi, R., & Wongwises, S. (2018). A comprehensive review on rheological behavior of mono and hybrid nanofluids: Effective parameters and predictive correlations. *International Journal of Heat and Mass Transfer*, *127*, 997–1012. https://doi.org/10.1016/j.ijheatmasstransfer.2018.07.103
6. Mooney, M. (1951). The viscosity of a concentrated suspension of spherical particles. *Journal of Colloid Science*, *6*, 162–170.
7. Brinkman, H. C. (1952). The viscosity of concentrated suspensions and solutions. *The Journal of Chemical Physics*, *20*, 571.
8. Thomas, D. G. (1965). Transport characteristics of suspension: VIII. A note on the viscosity of Newtonian suspensions of uniform spherical particles. *Journal of Colloid Science*, *20*, 267–77.
9. Lundgren, B. T. S. (1972). Slow flow through stationary random beds and suspensions of spheres. *Journal of Fluid Mechanics*, *51*, 273–299.
10. Batchelor, G. K. (1977). The effect of Brownian motion on the bulk stress in a suspension of spherical particles. *Journal of Fluid Mechanics*, *83*, 97–117.
11. Graham, A. L. (1981). On the viscosity of suspensions of solid spheres. *Applied Sciences Research*, *37*, 275–286.

12. Masoumi, N., Sohrabi, N., & Behzadmehr, A. (2009). A new model for calculating the effective viscosity of nanofluids. *Journal of Physics D: Applied Physics*, *42*, 1–6.
13. Gonçalves, I., Souza, R., Coutinho, G., Miranda, J., Moita, A., Pereira, J. E., et al. (2021). Thermal conductivity of nanofluids: A review on prediction models, controversies and challenges. *Applied Sciences*, *11*.
14. El-Magid Mohamed, M. A., Gutiérrez-Trashorras, A. J., & Meana-Fernández, A. (2024). Numerical study of rotating tube in tube heat exchanger using mono and hybrid nanofluids under transitional flow. *Processes*, *12*, 222.

6 Enhancement of Heat Transfer Using Hybrid Nanofluids

Convection is the main method responsible for heat transfer in fluids. The efficiency of this mechanism is primarily influenced by the thermophysical properties of the fluids. Moreover, enhancing the thermal conductivity of conventional fluids will greatly enhance the efficiency of the system. Thus, considering that solid materials have thermal conductivities that are far higher than those of normal fluids, the idea of introducing conducting particles into fluids was considered. Improving heat transmission has emerged as a crucial field of study to enhance the efficiency of different thermal systems. Hybrid nanofluids, which involve the combination of several nanoparticles, have emerged as a viable method to improve the efficiency of heat transfer. These nanoparticles possess distinct characteristics that can greatly enhance the efficiency of heat transfer. The distinctive characteristics of these features encompass heightened thermal conductivity, augmented convective heat transfer coefficient, and refined Nusselt number correlations. An advantage of utilizing hybrid nanofluids is their capacity to customize the thermal characteristics to meet individual needs. The customization of this tailoring can be accomplished by modifying the structure and amount of the nanoparticles present in the fluid. In addition, the combined use of various nanoparticles can result in enhanced thermal performance, surpassing that of nanofluids composed of a single component.

Hybrid nanofluids possess not just thermal properties but also provide benefits such as enhanced colloidal stability, decreased particle sedimentation, and improved flow characteristics. These qualities enhance its suitability for use in different thermal management systems, such as heat exchangers, cooling systems, and solar collectors. Moreover, the current investigation in this area seeks to enhance the production and durability of hybrid nanofluids, while comprehending their performance under varying temperature and pressure circumstances. By considering these factors, the practical utilization of hybrid nanofluids in heat transfer applications can be increased, resulting in more effective thermal systems with enhanced performance.

6.1 HEAT EXCHANGERS

Heat exchangers are devices that are capable of facilitating the passage of thermal energy between several fluids that have different temperatures. This trait makes heat exchangers efficient equipment for the transfer of heat.[1] Heat exchangers are utilized in a variety of industries, including but not limited to the production of power, the process, chemical, and food industries, the manufacturing industries, the electronics

DOI: 10.1201/9781003595137-6

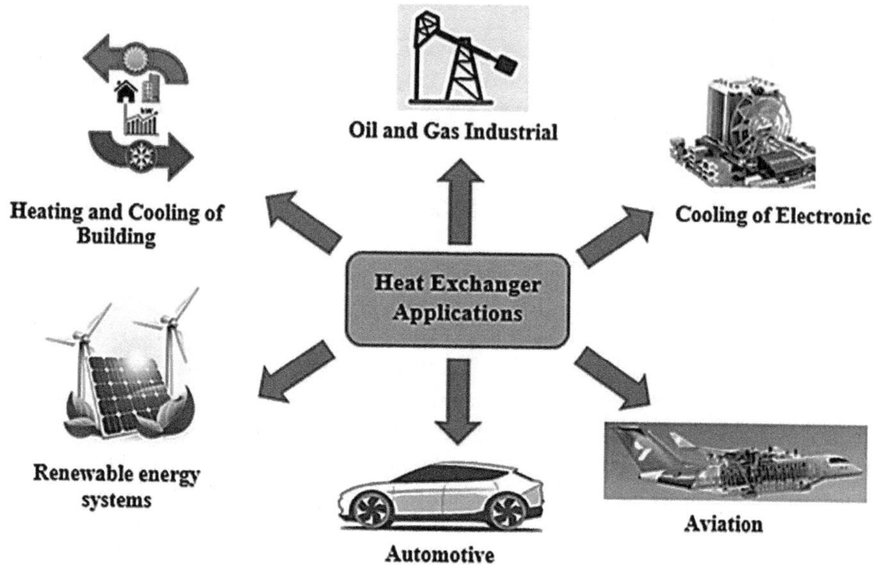

FIGURE 6.1 Applications of heat exchangers across a variety of industrial sectors.

industry, the air conditioning industry, the refrigeration industry, and space applications. The numerous applications of heat exchangers across a variety of sectors are depicted schematically in **Figure 6.1**.

It is usual practice in the automotive industry to locate them in the radiators of automobiles. Additionally, they are utilized in home appliances such as refrigerators, radiators, and coolers by the manufacturer. Heat exchangers are also utilized in the petrochemical and petroleum sectors to cool and preheat fluids.[2] In power plants, they are utilized to either reduce the temperature of the fluid that is used in the exhaust turbine or preheat the fluid that is used in the inlet boiler, which helps to lower the expenses associated with heating. A number of other applications of heat exchangers that are particularly noteworthy include solar collectors, the field of electronics, and the aeronautics industries.

Due to concerns about energy, space, and material conservation, as well as the global circular economy, there has been a rise in the efforts to produce more effective heat transfer equipment to cut expenses.[3] As a result of these efforts, the dimensions of heat exchangers have been diminished while preserving their thermal capacity. Thus, the primary thermal-hydraulic goals of a heat exchanger are to reduce and miniaturize the exchanger for a given heat load, enhance the capacity and operation of an existing exchanger, and reduce the pumping power. In recent decades, numerous studies have been conducted to improve the efficiency of heat transmission in heat exchangers and heat transfer fluids. Given this factor, some scholars investigated them by simplifying the heat exchangers to fundamental cavities. Several scholars have utilized these heat exchangers and designed them as open enclosures to study the fluid dynamics occurring within them. Scientists investigated the impact of the

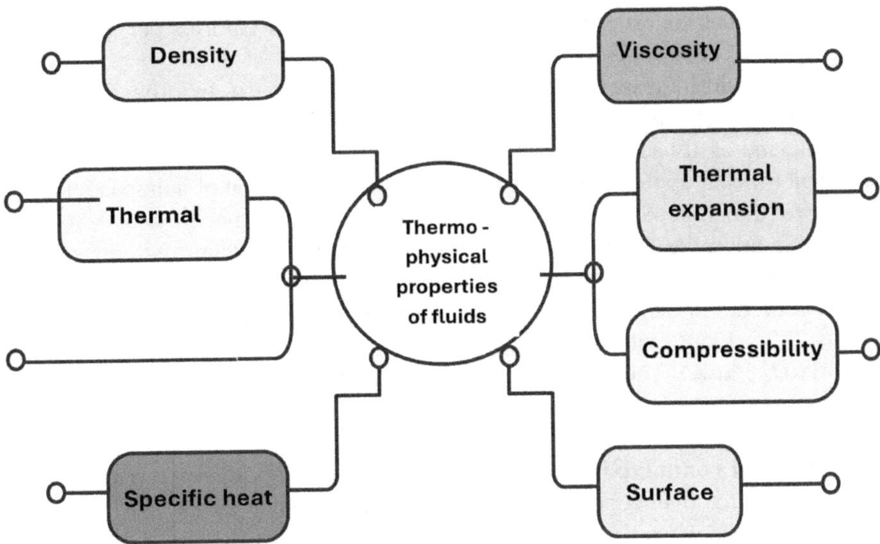

FIGURE 6.2 Thermophysical properties of fluids.

angle of inclination on the rate of heat transmission as part of their study. In addition to other techniques, altering the flow field has been achieved by introducing both a magnetic field and an electric field.[4]

The heat transfer properties are the function of the thermophysical properties of fluids. Thermophysical properties of fluids pertain to the physical attributes of substances in the fluid state, which are significant in determining their reaction to variations in temperature, pressure, and other environmental factors. These properties are essential in various applications, including efficient heat transfer, fluid dynamics, and process engineering. Common thermophysical characteristics of fluids are shown in **Figure 6.2**.

- **Density:** Density is the measure of mass per unit volume of a substance. Rho (ρ) is the usual symbol used to represent density, which is commonly measured in measures like kilograms per cubic meter (kg/m^3) or grams per cubic centimeter (g/cm^3). The density of a substance can change in response to changes in temperature and pressure. The density of nanofluid is examined in relation to temperature, as the density of fluids deviates from that of solids because the volume of the fluid depends on the temperature. In addition, nanofluids are formed by combining nanoparticles with base fluids, resulting in the presence of extra intermolecular interactions between them. The link between nanofluid concentration and temperature is measured to forecast the density of the nanofluid.
- **Specific heat capacity:** It is often known as specific heat, which refers to the quantity of heat energy needed to increase the temperature of a given mass of a substance by one degree Celsius (or Kelvin). The symbol used to represent specific heat capacity is "C_p". It is measured in measures such as

joules per kilogram per degree Celsius (J/kg°C) or calories per gram per degree Celsius (cal/g°C).

Due to its high specific heat capacity, water is extensively utilized for heat transfer purposes across various industries. Furthermore, water is affordable and easily accessible, and its thermal conductivity plays a vital role in heat transfer applications. Furthermore, the specific heat of nanoparticles is lower than that of water, resulting in a decrease in the specific heat of nanofluids upon the addition of nanoparticles. Moreover, when the concentration of nanofluid increases, a clear reduction in the specific heat capacity is observed. For example, the specific heat capacity of alumina nanoparticles is 779.2 J/kgK, which is significantly lower than that of water ($C_{p,\text{ water}}$ = 4180.6 J/kgK). Therefore, the inclusion of these nanoparticles at elevated concentrations has resulted in a decrease in the specific heat capacity of nanofluids.

- **Thermal conductivity**: Thermal conductivity refers to the capacity of a material to conduct heat. It measures the rate at which heat can go through a substance when there is a specific difference in temperature. The symbol used to represent thermal conductivity is "k". It is commonly expressed in measures such as watts per meter per degree Celsius (W/m°C) or calories per centimeter per second per degree Celsius (cal/cm sec °C).
- **Viscosity**: It refers to the degree of resistance that a fluid exhibits to deformation or flow. It measures the viscosity of a fluid as it moves. Viscosity is typically represented by the symbol μ and is quantified in units such as pascal-seconds (Pa·S) or poise (P). Fluids with low viscosity exhibit higher fluidity in comparison to fluids with high viscosity. As the fluid's viscosity increases, the amount of power required to pump it for transportation also increases since there is more resistance to its flow. Therefore, it is expected that there will be an increase in power usage. Given this understanding, numerous researchers have conducted experiments to investigate the impact of nanofluid concentration on the intermolecular interactions between particles, increasing the viscosity of the fluid.

  To mitigate the rise in viscosity and the associated pumping power requirements, it is preferable to use nanofluids with low concentrations (less than 1 Vol%) in various applications. The experimentation primarily focuses on investigating the impact of several factors such as nanofluid type, concentration, temperature, surfactant type, and concentration on the viscosity of the nanofluid.[5] In addition, numerical and mathematical equations have been derived from these experimental data to facilitate the prediction and forecasting of future data. Furthermore, these model equations can only be applied to specific nanofluids exclusively.
- **Thermal expansion coefficient**: It quantifies the extent to which the volume of a substance alters in response to variations in temperature. The thermal expansion coefficient, symbolized by β (beta), quantifies the ratio of volume change to temperature change. It is commonly measured in quantities of per degree Celsius (or per Kelvin).

- **Surface tension**: It refers to the inherent property of a liquid's surface to reduce its area, causing it to act like a flexible membrane. The phenomenon is a result of the cohesive forces among the molecules of the liquid. Surface tension is represented by the symbol σ (sigma) and is quantified in measures such as Newtons per meter (N/m) or dynes per centimeter (dyn/cm).
- **Compressibility**: It refers to the extent to which a substance's volume changes in response to a change in pressure. Beta (β) is commonly represented by the symbol β and is usually measured in units of per pascal.

Significantly, these properties vary among different fluids and are affected by operational parameters such as concentration, temperature, and pressure. The thermophysical characteristics of fluids are crucial in ensuring efficient heat transfer in applications. Since nanofluids consist of nanoparticles suspended in a base fluid, it is important to comprehend the thermophysical properties of these nanoparticles.[6] Within the set of thermophysical parameters, the thermal conductivity of nanoparticles is crucial for possible heat transfer applications. The combination of nanoparticles with high thermal conductivity is anticipated to enhance the thermophysical properties of the base fluid through a synergistic effect. In heat transfer applications, nanoparticles with excellent thermal conductivity are chosen.

6.1.1 Application of Nanofluids in Heat Exchangers

Heat exchangers are the devices which are used to transfer heat between the two fluids using temperature difference as a driving force. The fundamental concept of a heat exchanger is illustrated in **Figure 6.3**. As many of the applications involve heat transfer operation, many industries utilize these devices extensively. It won't be exaggerating to say that there are not many devices which are not utilizing these

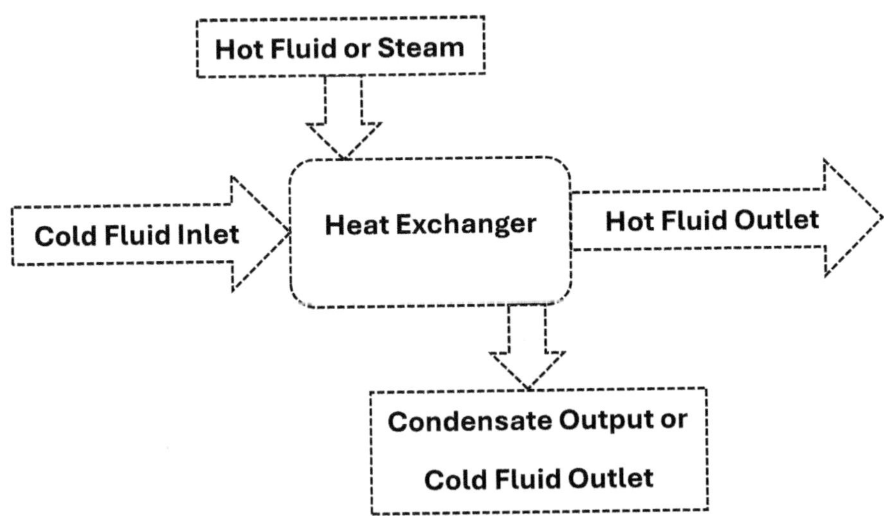

FIGURE 6.3 Fundamental working principle of a heat exchanger.

heat exchangers. To achieve maximum efficiency, heat exchangers are designed with an optimal wall surface area between two fluids. This design minimizes resistance to fluid flow via the exchangers, while also considering material cost limitations.[7] Integrating fins or corrugations into a heat exchanger has the capacity to enhance the effectiveness of heat transfer surfaces. These characteristics, which increase the total area, have the ability to control the movement of fluids or create turbulence.

An industrial heat exchanger is a device designed to transmit thermal energy between two or more mediums that have different temperatures. Industrial heat exchangers are employed in various areas, including power plant generation, the petroleum oil and gas industry, chemical processing plants, transportation, alternative fuels, cryogenic applications, air conditioning and refrigeration, and heat recovery. Furthermore, heat exchangers are found everywhere in the surroundings, as indicated by the existence of condensers, oil coolers, evaporators, air preheaters, and automotive radiators. In most heat exchangers, a heat transfer surface is used to separate the fluid. These heat exchangers utilize different flow designs to achieve the desired performance for different purposes. Heat exchangers can be classified in various ways. Industrial heat exchangers are commonly classified according to their construction, heat transfer mechanisms, surface compactness, flow arrangements, pass arrangements, phase of the process fluids, and pass arrangements, as shown in **Figure 6.4**.

Inadequate functioning of heat exchangers in any industry might lead to a complete system breakdown. Scientists were provided with a diverse range of methods

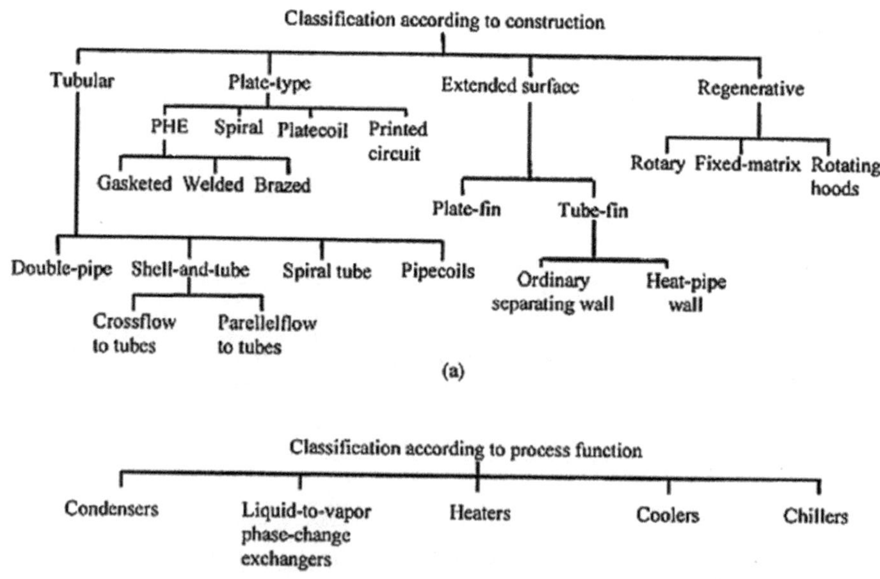

FIGURE 6.4 Different types of industrial heat exchangers based on the (a) design and (b) application.[8]

to address and solve these challenges. Utilizing nanofluids as a coolant in the heat exchanger without modifying its architecture is a more advantageous option when compared to other choices. That is to say, by increasing the thermal conductivity of the heat transfer fluid, the thermal efficiency of the heat exchanger can be improved without making any changes to the equipment. It is noteworthy that commonly used heat transfer fluids such as water, oils, and ethylene glycol have lower thermal conductivity compared to the majority of solids. It appeared more rational to include solid particles with strong thermal conductivity in the traditional fluid. Moreover, the research is commenced by dispersing particles of micron size. However, the study encountered problems such as decreased dispersion stability and the obstruction of microchannels, resulting in a limited enhancement in the thermal conductivity of the fluid. After nearly a century, researchers have begun utilizing various types of nanoparticles in the nanoscale region, resulting in significant enhancements in the thermophysical properties of heat transfer fluids. Nanofluids are fluids that contain particles in the nanoscale range. In recent years, there has been a rise in research focused on the synthesis and exploitation of sophisticated nanofluids with enhanced thermophysical properties in various types of industrial heat exchangers. This is driven by the need for greater global industry competitiveness and the recognition of the significant impact of energy on production costs.

The primary objective of the research on nanofluids in heat exchangers is to enhance the thermophysical properties of heat transfer fluids:

- Heat transfer fluids can be manufactured with enhanced thermal conductivity.
- Coolant (water) usage can be decreased in the heat exchanger.
- By decreasing the coolant quantity, the pumping power can be reduced so that the overall cost of the process.
- By minimizing the quantity of coolant in the heat exchanger, it becomes feasible to miniaturize the system.
- The utilization of miniaturized heat exchangers, such as those seen in vehicle radiator applications, leads to a decrease in the overall weight of the cooling system. The automobile and power generation sectors greatly benefit from this.
- For instance, this specific product, intended for car radiators, functions to prevent the boiling of water inside the radiator, which holds the engine cooling fluid. Moreover, it improves efficiency, hence reducing fuel consumption.
- Heat transfer fluids can be manufactured with enhanced thermal conductivity.

Typically, the heat transfer performance of a heat exchanger is evaluated by measuring the thermal conductivity and viscosity of nanofluids. This evaluation is done in terms of the heat transfer coefficient, Nusselt number, and rate of heat transfer. It is imperative to comprehend the impact of different parameters on enhancing the thermal conductivity of nanofluid, as depicted in **Figure 6.5**. All the properties exhibit a strong interconnection. In subsequent sections, the impact of these properties on thermal conductivity will be further elaborated.

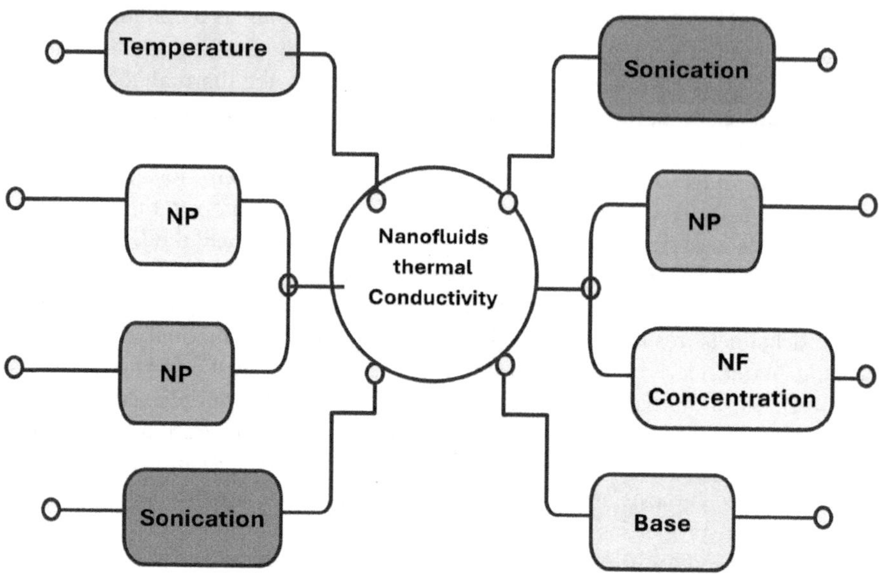

FIGURE 6.5 Effect of various properties on the thermal conductivity of nanofluids.

6.1.1.1 Effect of Nanoparticle Properties on the Thermal Conductivity of Nanofluids

Nanoparticle properties like nanoparticle type, size, shape, and concentration are important and are dependent on the nanoparticle. Different materials have different thermal conductivities as shown in **Table 6.1**. Generally, materials with higher thermal conductivities tend to enhance the overall thermal conductivity of the nanofluid. Common nanoparticle materials used in nanofluids include metals (such as silver, copper, and aluminum), metal oxides (such as alumina, titanium dioxide, and zinc oxide), and carbon-based materials (such as carbon nanotubes and graphene). It is critical to select a suitable nanoparticle for the application. This category includes elements such as nanoparticle size, shape, and thermophysical properties. One of the most essential issues is the role of electrical conductivity in nanoscience progress. When investigating the stability of nanofluids, it is critical to consider the morphology of the nanoparticles involved. On the contrary, the specific heat of nanoparticles is the factor that accounts for a variety of important properties. The level of protection just stated is crucial in terms of fostering and safeguarding thermal energy during transit and maintenance. The importance of nanoparticle density extends to a wide range of applications, including pumping. The majority of academic research has validated the notion that increasing the thermal conductivity of nanoparticles will result in an increase in the thermal conductivity of nanofluids. This general idea is one of the main reasons why nanofluids are used.

After selecting the appropriate nanoparticle, the selection of nanoparticle size plays an important role, as the size of nanoparticles significantly influences their thermal conductivity. Smaller nanoparticles typically have higher surface-to-volume ratios, which can lead to more efficient heat transfer. It is possible that smaller nanoparticles have greater interfacial surfaces per unit volume in comparison to bigger ones, which can result in an increase in the heat resistance between the nanoparticles. The increased

TABLE 6.1
Examples of Nanoparticle Type and Its Thermal Conductivity

Nanoparticle Type	Thermal Conductivity (*k*) in W/mk at 25°C
Graphene	3000
Ag	429
SiO_2	1.4
ZrO_2	2.2
MgO	48.4
TiO_2	8.4
ZnO	13
Al_2O_3	36
SiO_2	10.4
SiC	490
Al_2Cu	319
Ag_2Al	358
Al	204
Cu	383
Nanodiamond	1000
CuO	30
Multi-walled carbon nanotube	2000–3000

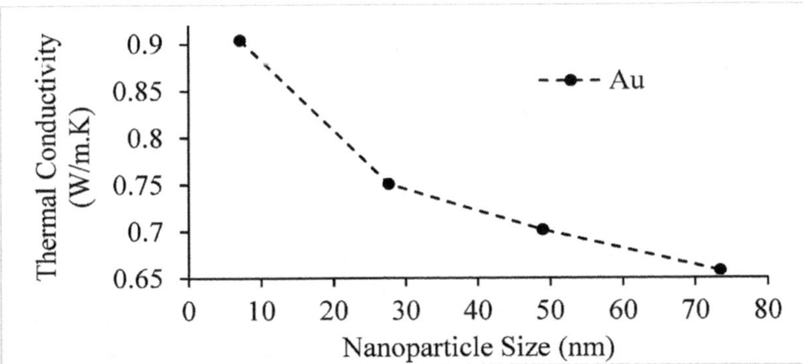

FIGURE 6.6 Effect of nanoparticle size on the thermal conductivity of 0.00026 Vol% of Au-water nanofluid.

surface area and Brownian motion effects, on the other hand, frequently dominate this impact and make it less noticeable. However, there is a critical size range below which the thermal conductivity enhancement diminishes due to increased interfacial thermal resistance. Smaller nanoparticles have higher specific surface areas in comparison to micron-size particles and even nanoparticles larger than 100 nm. Furthermore, as the size of the nanoparticles decreased, the level of contact between the nanoparticles and the molecules of the base fluid also increased. **Figure 6.6** validated the hypothesis that smaller-sized nanoparticles would enhance the thermal conductivity of nanofluids by virtue of their higher surface-to-volume ratio as the nanoparticles are in continuous Brownian motion.

Solve:
How does surface area get larger compared to the volume of the cell?

Solution:
Let's assume the cell is in spherical shape.

$$\text{Surface area of sphere} = SA = 4\pi r^2$$

$$\text{Volume of sphere} = V = \frac{4}{3}\pi r^3$$

Now assume the cell radius is 1 and 10 cm.

Surface-to-volume ratio:

$$\frac{\text{Surface area of sphere}}{\text{Volume}} = \frac{4\pi r^2}{\frac{4}{3}\pi r^3} = \frac{3}{r}$$

Case 1: Cell radius is 1 cm.

$$\frac{\text{Surface area of sphere}}{\text{Volume}} = \frac{4\pi r^2}{\frac{4}{3}\pi r^3} = \frac{3}{r} = \frac{3}{1} = 3$$

Case 2: Cell radius is 10 cm.

$$\frac{\text{Surface area of sphere}}{\text{Volume}} = \frac{4\pi r^2}{\frac{4}{3}\pi r^3} = \frac{3}{r} = \frac{3}{10} = 0.3$$

This shows that the lower the particle size higher the surface-to-volume ratio.

Consequently, the utilization of smaller nanoparticles has the capacity to enhance heat transfer by augmenting the interface area between the nanofluid and the surface that is being either heated or cooled by the nanofluid. Moreover, the larger the size of the particle, the more prone it is to integrate with other particles and settle due to agglomeration. Due to the decreased stability of the particles in the base fluid, instead of being completely dispersed, a two-layered pattern is observed as a result of particle sedimentation. This behavior can cause significant issues with pipe blockage, as the particles settle in the bottom, resulting in a decrease in the overall improvement of the nanofluid's thermal conductivity.

Typically, to reduce the aggregating and settling of nanoparticles, a reliable stability mechanism is employed. This involves altering the surface properties of the nanoparticles by processes such as functionalization, pH adjustment, surfactant addition, and ultrasonication. In simple terms, these approaches enhance the Brownian motion of the particles. The smaller the particle size, the greater the absorption of kinetic energy and the more efficient the Brownian motion of the particles. This, in turn, greatly enhances the dispersion of nanoparticles and synergistically enhances the thermophysical properties of nanofluid.

6.1.2 Important Points to Remember

- Typically, when solid particles are added to a liquid, they absorb ions from the solution and develop a surface charge, depending on the pH of the solution. The adsorption of ions onto the surface of nanoparticles creates a potential difference known as zetapotential (represented by ζ and measured in millivolts (mV)) between the solid particles and the liquid.
- In simple terms, a higher zetapotential value indicates stronger electrostatic repulsion forces between the two phases and greater stability of the nanofluid dispersion. Since the stability of the nanofluid directly affects its heat transfer properties, understanding zetapotential is crucial.
- Particle size analyzers are commonly used to measure the zetapotential value of nanofluids.
- When dealing with measurements in the nanoscale range, it is important to take into account the effect of quantum confinement due to size. The phenomenon of quantum confinement size effect occurs when the size of a particle is smaller than the wavelength of an electron.
- Reducing the size of nanoparticles leads to a large increase in the dominance of quantum effects, which in turn affects the thermophysical properties.
- Furthermore, it is crucial to optimize the dimensions of nanoparticles and ensure a consistent monochromatic distribution of particle sizes.
- The particle size distribution analysis is common practice in understanding the dispersion mechanism of nanofluids would facilitate the optimization of thermal conductivity enhancement in specific applications.

Along with particle type and size, nanoparticle shape affects the thermal conductivity of nanofluids. **Figure 6.7** shows various forms and geometries of natural or engineered nanoparticles for reference. It shows that it is not necessary to assume the nanoparticle shape to be spherical always. So, based on the shape of the particle, the surface-to-volume ratio is varied. Moreover, precisely manipulating the form can pose a challenge. The shape of nanoparticles generated can be influenced by factors such as reaction kinetics, type and concentration of precursor, temperature, pressure, and surfactants. Attaining consistency in form throughout a substantial quantity introduces an additional level of intricacy. Further, extensive studies have shown that nanoparticles with spherical shapes have shown better dispersion stability and fewer issues with clogging of microchannels and pipes. Another important point is that the synthesis of any particle with a shape other than spherical shows a higher cost burden.

It has been shown that nanoparticles with shapes other than spherical have a lower settling velocity due to their higher surface areas. The drag force exerted on an object moving through a fluid increases in direct proportion to the object's surface

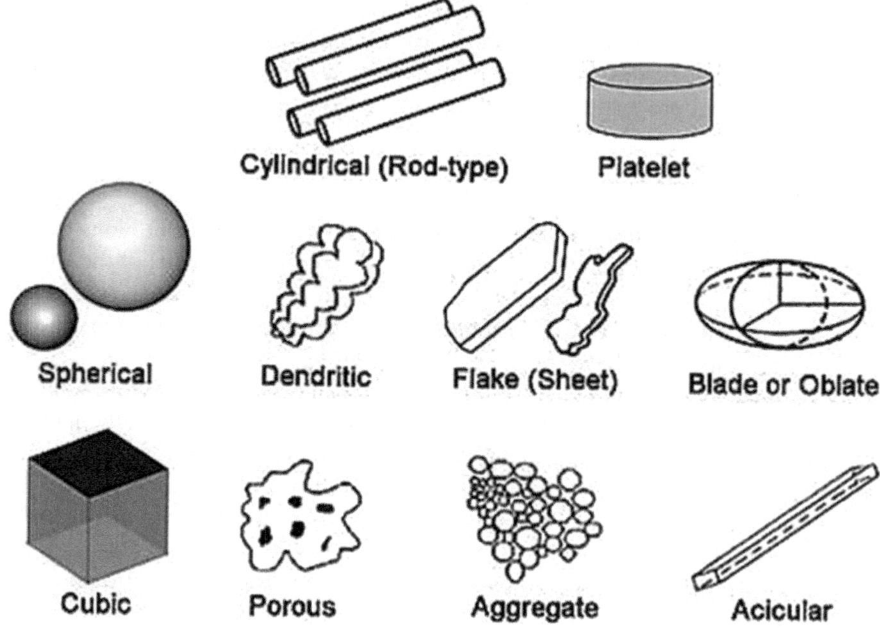

FIGURE 6.7 Examples of several forms of nanoparticles.[9]

area. This holds true for particles with non-uniform shapes. However, the impact of this phenomenon may vary when considering particles with elongated shapes such as nanofibers, nanowires, nanotubes, and nanorods, which possess a variable aspect ratio compared to spherical particles. The reason for this is that elongated nanoparticles encounter reduced resistance while moving parallel to their long axis as opposed to moving perpendicular to their long axis.

Nanoparticles with shapes that are irregular exhibit varying enhancement rates along different axes of thermal conductivity compared to spherical particles. The variance in effective thermal conductivity enhancement may be attributed to the surface-to-volume ratio, also known as the aspect ratio. As a result, there were several inconsistencies in the findings of studies about the impact of nanoparticle form on the effective thermal conductivity of nanofluids. This phenomenon is particularly evident in the case of hybrid nanofluids, which contain nanoparticles with varying sizes and shapes. Furthermore, while dealing with hybrid nanofluids, as it's a combination of more than one nanoparticle, if the nanoparticles are in different shapes the system becomes more complicated in achieving the stable suspension.

Once the nanoparticle's type, size, and shape have been determined, optimizing their concentration becomes a crucial factor. The thermal conductivity of nanofluid is significantly increased due to the addition of nanoparticles, which have a greater thermal conductivity than the base fluid.[10] This enhancement is achieved by a synergistic effect. The majority of investigations have observed a linear rise in effective thermal conductivity as the concentration of nanofluids increases. In a stable suspension, the nanoparticles exhibit complete Brownian motion. The presence of a higher number of particles leads to an increased occurrence of thermal bridging phenomena, hence enhancing the relative characteristics of the resulting fluid. However, contrary to the findings, other writers have posited the possibility of particle agglomeration occurring at elevated concentrations of nanoparticles. In this scenario, we can anticipate hindered settling instead of free settling.

6.1.3 Free Settling

- Free settling occurs when particles, located at a significant distance from each other, begin to settle through a fluid media alone due to the force of gravity.
- The particles travel independently from each other due to their significant distance apart.
- Therefore, the settling velocity depends on various variables of the nanoparticles, such as their type, shape, size, and density, as well as the viscosity and temperature of the fluid.
- When the Reynolds number (N_{Re}) is less than 1, the settling velocity of a particle can be measured using the simple Stokes law in the case of free settling.

$$\text{Settling velocity} = \frac{g D_P^2 \left(\rho_p - \rho\right)}{18\mu} \tag{6.1}$$

Here, g = Acceleration due to Gravity (9.81 m/s^2)
D_p = Diameter of the particle
ρ_p = Density of the particle (kg/m^3)
ρ = Density of the fluid (kg/m^3)
μ = Viscosity of the fluid

6.1.4 Hindered Settling

- Hindered settling happens when there is a high concentration of particles in a fluid. For instance, when there are large concentrations, the particles are closer together.
- The proximity of particles to one other results in the modification of particle settling velocities due to collisions with other particles or the wall surface.
- Collisions between particles result in an increase in the applied drag force. Hindered settling velocities are less than free settling velocities.
- Overcrowding can result in the aggregation and flocculation of particles.
- The hindered settling velocity is greatly influenced by the concentration of nanoparticles.

6.1.5 Important Points

1. While the thermal conductivity of nanofluid is greatly influenced by the concentration of nanoparticles, it is not advisable to continuously add

nanoparticles to the base fluid without considering other factors. By raising the concentration of nanoparticles, the cost of the heat transfer coolant will also increase, as nanoparticles are not inexpensive. Furthermore, it incites the interaction and clustering of particles.
2. Agglomeration is the process of nanoparticles coming together and forming clusters or aggregates. This might result in a decrease in their effective surface area and inhibit the transmission of heat.
3. These clusters may precipitate more rapidly. Sedimentation is the process by which nanoparticles settle out of a liquid suspension under the influence of gravity.
4. Sedimentation can cause an uneven dispersion of nanoparticles and decrease heat conductivity in specific areas of the nanofluid.
5. Lack of consistency in the outcomes. This factor renders it unreliable and diminishes the potential for industrial scalability.
6. Therefore, it is crucial to optimize the concentration of nanofluid considering all the influencing parameters.

As soon as the parameters of nanoparticles have been comprehended, the selection of an appropriate base fluid becomes a critical consideration. Nanofluids have a viscosity that is very dependent on the type of fluid that is being used during their production. Some examples of base fluids that are often used include ethylene glycol, which has a viscosity of 18.376 cP, which is significantly higher than the viscosity of water, which is 0.890 cP. As the viscosity of the fluid grows, it eventually becomes necessary to make modifications to the performance of the pump to compensate for the increased resistance to shear. Therefore, it can be observed that there is a slight reduction in the flow rate, a more noticeable reduction in the head or pressure, and a significant increase in the amount of power that is consumed. As a consequence of this, the viscosity of the fluid has a significant relationship with the size of the pumping power, which is independent of the amount of shear stress that is present.

Additionally, the type of fluid that is used is a significant factor that plays a role in determining the thermal conductivity of the nanofluid. There is a correlation between the thermal conductivity of the base fluid and the thermal conductivity of the nanofluid. It is important to note that the choice of the right fluid is dependent on the particular use of nanofluid inside the system. Without a doubt, water with a low viscosity stands out as the most prominent and widely used fluid. This is due to the fact that it is readily available in large quantities and possesses superior thermal properties in contrast to other fluids.

6.2 SUMMARY

- Nanofluids are being marketed as heat transfer fluids due to their lower thermal conductivity compared to solids. The addition of nanoparticles with higher thermal conductivity into any prospective coolant has a synergistic effect on the heat transfer capabilities.
- Instead of making changes to the heat exchanger, it is more cost-effective to implement a highly efficient cooling system.

- By increasing the pace at which the coolant is recirculated, it is possible to decrease the amount of coolant needed.
- By reducing the water quantity, it is possible to optimize pumping costs and overall process costs.
- By reducing the amount of coolant in the heat exchanger, it becomes possible to make the system smaller in size.
- Implementing compact heat exchangers, similar to those found in automotive radiator systems, results in a reduction in the total weight of the cooling system. This has a significant positive impact on the vehicle and electricity generation industries.
- A thorough comprehension of the several characteristics affecting the thermal conductivity of nanofluids is crucial for maximizing their efficiency in numerous applications, such as heat transfer fluids, thermal management systems, and advanced cooling technologies.

REFERENCES

1. Sonawane, S. S., Thakur, P. P., Malika, M., & Ali, H. M. (2023). Recent advances in the applications of green synthesized nanoparticle based nanofluids for the environmental remediation. *Current Pharmaceutical Biotechnology*, *24*(1), 188–198.
2. Malika, M., & Sonawane, S. S. (2022). The sono-photocatalytic performance of a Fe2O3 coated TiO2 based hybrid nanofluid under visible light via RSM. *Colloids and Surfaces A: Physicochemical and Engineering Aspects*, *641*, 128545.
3. Malika, M., & Sonawane, S. S. (2022). MSG extraction using silicon carbide-based emulsion nanofluid membrane: Desirability and RSM optimisation. *Colloids and Surfaces A: Physicochemical and Engineering Aspects*, *651*, 129594.
4. Malika, M., Jhadav, P. G., Parate, V. R., & Sonawane, S. S. (2023). Synthesis of magnetite nanoparticle from potato peel extract: Its nanofluid applications and life cycle analysis. *Chemical Papers*, *77*(2), 1081–1094.
5. Malika, M., Pargaonkar, A., & Sonawane, S. S. (2023). Experimental and statistical analysis of MWCNT hybrid nanofluid-based multi-functional drilling fluid. *Chemical Papers*, *77*(11), 6773–6784.
6. Malika, M., & Sonawane, S. S. (2024). Ecological optimization and LCA of TiO2-SiC/water hybrid nanofluid in a shell and tube heat exchanger by ANN. *Proceedings of the Institution of Mechanical Engineers, Part E: Journal of Process Mechanical Engineering*, *238*(1), 45–55.
7. Malika, M., Pargaonkar, A., & Sonawane, S. S. (2024). Performance of an emulsion nanofluid membrane for the extraction of antimony heavy metal: Experimental and numerical investigation. *Journal of Sustainable Metallurgy*, 1–12.
8. Shah, R. K., Thonon, B., & Benforado, D. M. (2000). Opportunities for heat exchanger applications in environmental systems. *Applied Thermal Engineering*, *20*(7), 631–650. https://doi.org/10.1016/S1359-4311(99)00045-9
9. Taha-Tijerina, J. J. (2018). Thermal transport and challenges on nanofluids performance. In *Microfluidics and Nanofluidics*. InTech. https://doi.org/10.5772/intechopen.72505.
10. Malika, M., & Sonawane, S. (2024). A review on the application of nanofluids in enhanced oil recovery. *Current Nanoscience*, *20*(3), 328–338.

7 The Application of Hybrid Nanofluids in the Field of Wastewater Treatment

Water is recognized as the most crucial natural resource that is in great demand by humanity. Nevertheless, there remains a constraint on the availability of high-quality and easily transportable potable water. Even though living organisms rely on water for their life, the majority of water usage worldwide is attributed to industrial and agricultural activities. Compared to the operations of industry and agricultural output, the water consumption by living creatures is negligible. Water is a necessary raw ingredient for several activities such as beverage manufacture, vehicle manufacturing, crude oil separation, farmland irrigation, and more. Water is utilized in a diverse range of companies and businesses. Due to these circumstances, water becomes polluted as it is used in many processes, each of which might lead to a specific type of water contamination.

Despite the abundance of water resources on Earth, it is not acceptable to use water in factories and dispose of it elsewhere without proper treatment. The contamination of the water would propagate pathogenic microorganisms, leading to the spread of diseases. Hence, it is imperative to seek a viable and enduring resolution that effectively addresses the multitude of obstacles associated with the elimination of the substantial concentration of pollutants found in the water. A variety of chemicals were employed in the disinfection of wastewater.

7.1 WASTEWATER TREATMENT METHODS

Wastewater treatment methods can be categorized into three main groups as shown in **Figure 7.1**:

1. Physical or primary treatment
2. Chemical or secondary treatment
3. Biological or tertiary treatment

7.1.1 Physical Treatment

Primarily, the treatment of wastewater using physical processes focuses on separating solids from liquids, with filtration playing a crucial role in this regard. Screening and pumping are the first steps. The wastewater that is being brought in is screened

DOI: 10.1201/9781003595137-7

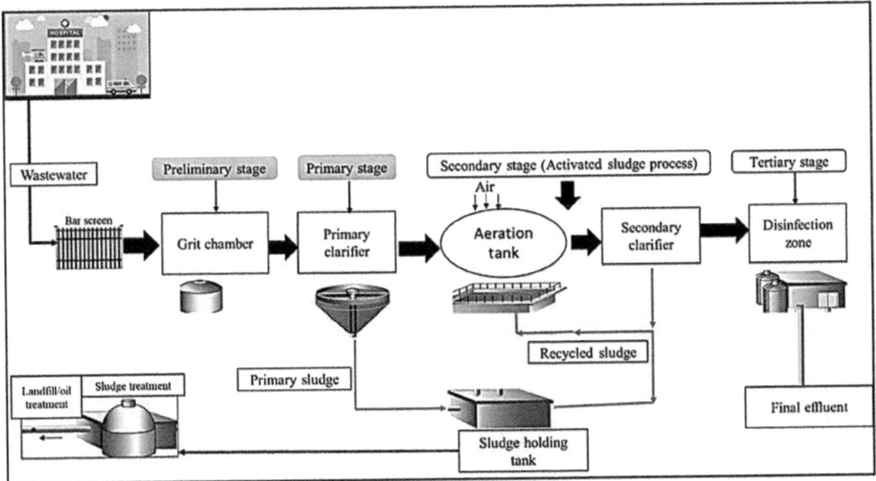

FIGURE 7.1 A simple guide to wastewater treatment and the various steps involved.[1]

by screening equipment, which removes debris such as rags, pieces of wood, plastics, and grease from the water. Following the removal of the material, it is washed, compressed, and then disposed of in a landfill. After the wastewater has been screened, it is pumped into the subsequent step, which is the removal of grit.

Heavy but fine materials, such as sand and gravel, are extracted from the wastewater during this stage of the cleansing process. A landfill is another location where this material is disposed of. The material, which will settle, but at a slower rate than step two, is removed by means of large circular tanks that are referred to as clarifiers. The material that has settled, which is referred to as primary sludge, is pumped off the bottom of the tank, and the wastewater is discharged from the top of the tank. Floating debris, such as grease, is removed from the top of the mixture and transported to digesters along with the material that has settled. In addition, chemicals are added during this stage to eliminate phosphorus.

7.1.2 Chemical Treatment

This method relies on chemical interactions between the contaminants and the person applying the chemicals. These methods provide aid in either fully eliminating contaminants from water or counteracting the detrimental effects associated with contaminants. Chemical treatment methods can be utilized either on their own or as part of a treatment process that also involves physical processes. Both of these applications are feasible.

7.1.3 Biological Treatment

Most of the wastewater treatment occurs during this stage. During biological degradation, microorganisms consume pollutants, leading to the conversion of the pollutants into nitrogen, water, and cell tissue. Analogous to the biological processes occurring in the depths of lakes and rivers, the degradation processes in these regions require several years to reach completion. Nevertheless, the biological activity occurring in this step is highly comparable. During this phase, the processed wastewater is isolated from the organic matter in the aeration tanks using large circular tanks called secondary clarifiers. As a consequence, the outcome is the creation of an effluent that has undergone treatment to a degree exceeding 90%. The activated sludge, which is the biological material, is gathered from the clarifiers' bottom and transported back to the aeration tanks in a continuous manner.

Biological treatment of wastewater may seem straightforward due to its reliance on natural processes to aid in the decomposition of organic compounds. Nevertheless, in actuality, it is an intricate procedure that remains partially comprehended and occurs at the convergence of biology and biochemistry. Organic matter, such as garbage, organic wastes, partially digested foods, heavy metals, and toxic substances, can all be present in wastewater. Biological treatments for decomposing organic matter commonly utilize microorganisms, including bacteria, nematodes, and other small organisms.

7.2 DISINFECTION OF WASTEWATER

The disinfection of wastewater is a crucial step in the wastewater treatment process to ensure that harmful pathogens, such as bacteria, viruses, and parasites, are completely removed before the treated water is released into the environment or reused for various purposes. Various techniques are commonly employed to sterilize wastewater, such as:

i. **Chlorination** is a widely used method for disinfecting wastewater, and it is one of the most commonly employed techniques. Chlorine is commonly added to wastewater using chlorine gas (Cl_2), sodium hypochlorite (NaOCl), or calcium hypochlorite (Ca(OCl)), among other techniques. Chlorine reacts with organic molecules and microbial infections, causing their cellular structures to be damaged and rendered inert. However, chlorination can lead to the formation of hazardous disinfection byproducts (DBPs), such as trihalomethanes and haloacetic acids. These DBPs are regulated due to the potential hazards they provide to human health.

ii. **Chlorine dioxide:** Chlorine dioxide, or ClO_2, is an effective alternative disinfectant that is highly effective against a wide range of illnesses and produces fewer DBPs compared to chlorine. Chlorine dioxide can be used alone or in combination with chlorine to disinfect wastewater.

iii. **Ozonation**: It is a method of cleaning wastewater by exposing it to ozone (O_3), a powerful oxidizing agent. Ozone is a potent pathogen killer because it oxidizes the biological components of illnesses. Additionally, it aids in the breakdown of organic substances and the decrease of unpleasant smell and discoloration in wastewater. Ozonation is a more environmentally friendly production process than chlorination since it does not generate chlorinated DBPs.

iv. Fourthly, **ultraviolet (UV) irradiation** is a chemical-free form of disinfection. It utilizes UV light to deactivate bacteria by inducing damage to their DNA or RNA. UV disinfection systems offer ecological benefits and are compact and energy efficient. However, the effectiveness of UV disinfection can be affected by several factors, such as the cloudiness of the water and the existence of organic substances.
v. **Advanced oxidation processes (AOPs)** represent the fifth procedure. AOPs utilize highly reactive hydroxyl radicals (•OH) to oxidize and eliminate organic contaminants and microbiological pathogens found in wastewater. Hydrogen peroxide (H_2O_2), when exposed to UV radiation or ozone, can effectively purify wastewater and eliminate small amounts of contaminants. This technique is known as AOPs.
vi. **Membrane filtration** encompasses several technologies such as microfiltration, ultrafiltration, and nanofiltration. These techniques effectively eliminate pathogens from wastewater by forcing it through membranes with certain pore sizes. Although membrane filtration is not a standalone disinfection method, it can be combined with other disinfection methods due to its ability to form a physical barrier against microbiological pollutants.

7.2.1 Limitations of Conventional Methods

According to **Figure 7.2**, there are global priorities assigned to different techniques. The experts prefer UV radiation combined with advanced oxidation technology the most (28.38%), followed by the natural systems alternative.

However, once again, the presence of harmful substances had a negative effect on the growth of the bacteria. Multiple foodborne diseases and waterborne microbes were negatively impacting human health. Due to the growing resistance of bacteria to

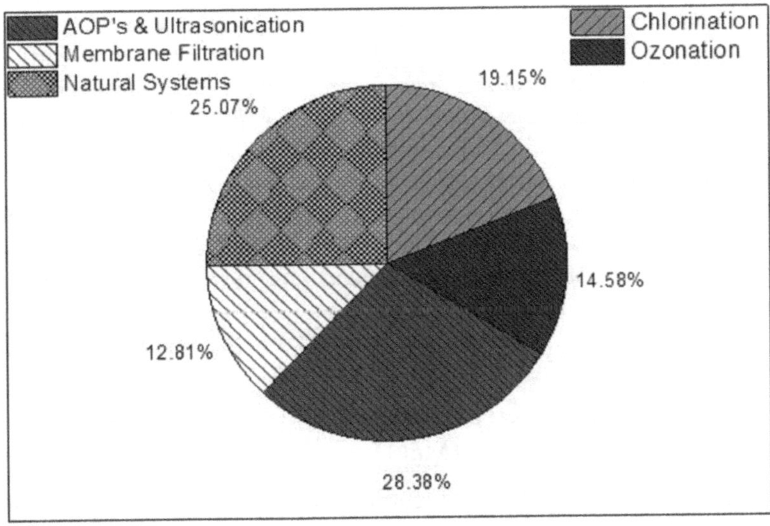

FIGURE 7.2 Considerations for prioritizing disinfection technologies on a global scale.

commonly used antibiotics and disinfectants, multiple studies have been conducted to enhance antimicrobial techniques. In the field of antimicrobial nanoparticles, numerous studies demonstrating their beneficial effects have been published in recent years. The utilization of metal and metal oxide nanoparticles may be a viable substitute for specific antibacterial therapies. Antimicrobial nanoparticles have the potential to be employed in the pharmaceutical and biomedical sectors to sterilize medical devices. Furthermore, these nanostructures have the capacity to be employed in the production of chemical disinfectants, as well as in various applications such as coatings and food preparation operations. Studies have shown that metal nanoparticles, especially those composed of metal oxides, have significant antibacterial properties. Metal oxide nanoparticles exhibit exceptional antibacterial properties owing to their remarkable thermal stability and superior efficacy against drug-resistant microbial infections. The combination of these two qualities renders them a remarkably effective antibacterial agent. However, the usage of specific metal oxide nanoparticles is restricted due to their heightened toxicity at larger concentrations.

Moreover, it has been reported that utilizing functionalization, ion doping, and polymer conjugates of these nanoparticles could be beneficial in reducing the toxicity of these nanoparticles. The main goal is to create cost-effective and efficient inorganic antimicrobial compounds as a viable alternative to the current antibiotics. Illustrations of such agents encompass nanoparticles made of metal and metal oxide. This innovative creation demonstrates the promising prospects that lie ahead for the pharmaceutical and medical sectors. Bacterial adhesion to human tissue surfaces involves several important physical interactions, including Brownian motion, van der Waals attraction forces, surface electrostatic charge, gravitational forces, hydrogen bonding, ionic contacts, hydrophobic interactions, and electric dipole moment.

Recent studies and examinations have shown that zinc oxide nanoparticles (NPs) have the ability to inhibit the spread of diseases. Furthermore, animal dung is recognized to include a diverse range of bacteria, some of which have the potential to cause disease in both animals and people, while others are harmless. Foodborne pathogens such as *Enterococcus faecalis*, *Proteus vulgaris*, *Pseudomonas aeruginosa*, *Serratia marcescens*, *Salmonella typhi*, *Staphylococcus epidermidis*, methicillin-resistant *Staphylococcus aureus*, *Bacillus anthracis*, *Bacillus cereus*, and *Streptococcus pyogenes* can cause various infections including hospital infection, endocarditis bacteremia, urinary infection, genital duct infection, gastroenteritis, pneumonia, septicemia, respiratory and gastroenteritis infections, and sore throat. *S. aureus* is also associated with skin and urinary infections. The selected community continues to suffer a recurring and major problem of invasion by gangrene, scarlet fever, and acute glomerulonephritis due to a lack of understanding. The implementation of potent novel antimicrobial drugs is crucial for managing pathogenic microorganisms, particularly those that are resistant to antibiotics.

7.3 EXPLORING THE POTENTIAL OF NANOTECHNOLOGY IN WASTEWATER TREATMENT

The efficacy of conventional water treatment methods in removing contaminants such as metals and microorganisms from water is often limited. Another concern arises from the generation of DBPs, which possess the capacity to pose risks to human health. The chemical disinfectants undergo a reaction with both organic matter and inorganic ions present in the water, leading to the creation of DPBs. Available methods of disinfection include the addition of chemicals (such as chlorine and iodine) as well as the utilization of UV light, ozone, radiation, and other means. Within the context of water and wastewater treatment plants, chlorine and the compounds that are related to it are by far the most commonly used disinfectants. On the other hand, it was discovered in the early 1970s that the utilization of free chlorine, chloramines, and ozone results in the production of a number of DBPs that are unpleasant and potentially hazardous. In the published research, there have been over 600 different DBPs reported. Over the course of several years, researchers have been searching for alternative and innovative methods of disinfection to address the concerns regarding the formation of DBPs.

The use of nanoparticles in disinfection-related industries, such as wastewater treatment plants, has gained significant attention due to the rapid growth and application of nanotechnology. Nanoscale materials, defined as materials with at least one dimension smaller than approximately 100 nm, exhibit unique electrical, thermal, mechanical, chemical, and biological characteristics when engineered appropriately. The small size and large surface area of nanomaterials both contribute to their high adsorption and reactive properties. Moreover, there have been reports indicating that nanomaterials demonstrate a significant level of mobility in relation to solutions. Through the application of diverse nanomaterials, it has been documented that heavy metals, organic pollutants, inorganic anions, and bacteria can all be eradicated. A wide range of nanomaterials, including zero-valent metal nanoparticles, metal oxide nanoparticles, carbon nanotubes (CNTs), and nanocomposites, have undergone a thorough investigation to assess their potential uses in water and wastewater treatment. Recently, research and investigation have focused on the antimicrobial properties of different nanoparticles in the dispersion form (nanofluids), including silver, titanium dioxide, zinc oxide, magnesium oxide, calcium oxide, and silicon dioxide.

Nanofluids are a type of fluids that consist of a base fluid containing nanoparticles with sizes ranging from 1 to 100 nm. Ethylene glycol and water-based nanofluids have been widely utilized as the predominant nanofluids in recent decades. The thermo-physical properties of these functionalized smart nanofluids have shown a substantial enhancement. Given our current understanding of the behavior of large quantities of materials, we cannot accurately forecast the behavior of nanofluids. Nanofluids possess distinct values for properties such as thermal conductivity, thermal diffusivity, viscosity, and convective heat transfer coefficients, which differ from those of base fluids. Furthermore, nanofluids are employed in the biomedical sector for various applications such as magnetic cell separation, drug delivery, temperature elevation, and improved contrast in magnetic resonance imaging. Currently, researchers are shifting their focus toward exploring additional facets of nanofluids, particularly their potential applications in environmental domains. Recent reports indicate that nanofluids possess augmented antimicrobial properties. Studies have shown that TiO_2, ZnO, nC_{60}, magnesium oxide, and CNTs exhibit remarkable antibacterial properties. Simultaneously with the development of nanofluids, there is an increasing interest in using nanomaterials for the disinfection step of wastewater treatment.

7.4 TYPES OF BACTERIA

Bacteria are unicellular microorganisms that are not visible to the naked eye. Bacteria are omnipresent, predominantly independent organisms, typically comprising a single biological cell. Prokaryotic microorganisms make up a significant domain. Bacteria, which are usually a few micrometers (μm) long, were one of the earliest life forms to emerge on Earth and may be found in most of its environments. Bacteria can be found in both the internal and external environments of creatures, including humans. Bacteria are ubiquitous in various environments, including surfaces, water, soil, and food, and play a crucial role in Earth's ecosystems. According to the National Human Genome Research Institute, although certain bacteria can cause illnesses and be detrimental to people, the majority of bacteria are benign. Even our human body requires specific strains of bacteria, particularly those residing in the gastrointestinal tract, to carry out essential physiological processes.

All creatures consist of fundamental units known as cells. A cell is a bounded structure with a distinct outer layer called the plasma membrane. The presence of the living state is dependent on the biological membrane, which is a characteristic shared by all living organisms. Bacteria exhibit a simple biological structure, including uncomplicated components such as the cell wall, cell membrane, cytoplasm, nuclear material, plasmid, ribosomes, and other elements (**Figure 7.3**).

The bacterial cell envelope, which consists of the membrane(s) and other structures that enclose and safeguard the cytoplasm, is far from being a mere membrane. Unlike the cells found in more complex creatures, bacteria are exposed to uncertain, weak, and frequently unfriendly surroundings. Bacteria have developed an intricate and intricate cell envelope to ensure their survival. This envelope serves to shield them while still enabling the selective movement of nutrients from the external environment and waste products from the internal environment. The subsequent

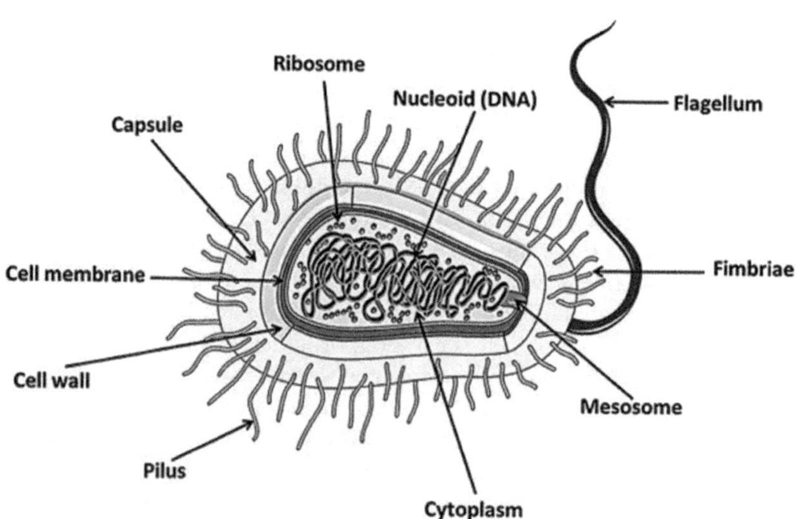

FIGURE 7.3 A broad overview of bacteria cell anatomy.[2]

discourse pertains to the arrangement, constitution, and roles of the diverse layers and compartments that constitute this extraordinary cellular structure. It is readily understood that a living system cannot function without the capacity to create distinct compartments where its components are separated. Specialized functions are localized within specific compartments due to the presence of specific types of molecules that are limited to those compartments. Nevertheless, membranes have a more complex function than just separating various molecules.

Bacteria are categorized as gram-positive and gram-negative based on variations in their structure and lipid composition of the cell wall. The various varieties are differentiated based on their susceptibility to staining using the staining technique pioneered by Danish researcher Dr. Hans Gram in 1884. The primary distinction between the two groups is in the absence of an external membrane in the gram-positive cellular envelope, as shown in **Figure 7.4**. Classification of bacteria based on the lipid structure has been listed in **Table 7.1**.

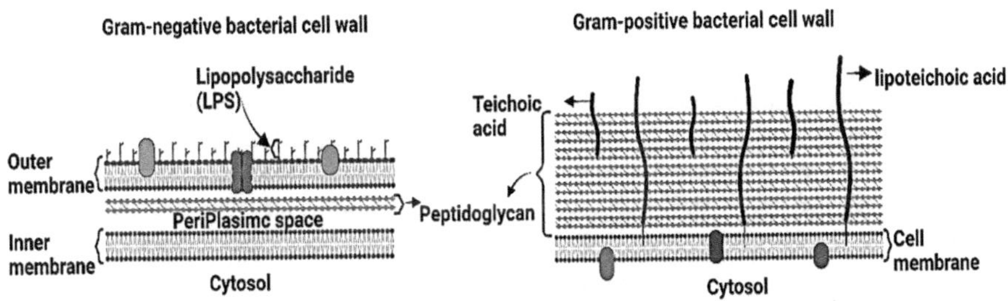

FIGURE 7.4 The fundamental structural distinctions between gram-positive and gram-negative bacteria.[3]

TABLE 7.1
Examples for Gram-Positive and Gram-Negative Bacteria

Gram-Positive Bacteria	Gram-Negative Bacteria
Staphylococcus aureus	*Salmonella typhi*
Staphylococcus epidermidis	*Salmonella typhimurium*
Staphylococcus saprophyticus	*Pseudomonas aeroginosa*
Streptococcus pyogenes	*Pseudomonas flourscences*
Streptococcus pneumonia	*Proteus vulgaris*
Streptococcus mutanus	*Proteus merabilis*
Streptococcus feacalis	*Klebsiella pneumonia*
Streptococcus aglagtia	*Enterobacter aerugenes*
Lactic acid bacteria (yogurt)	*Acinetobacter*
Bifidobacterium	*Escherichia coli* (*E. coli*)
Mycobacterium tuberculosis	*Serratia* spp.
Anthrax	Acetic acid bacteria (vinegar)
Enterococci	*Xanthomonas* (Xanthan gum)
Hemolytic bacteria	*Zymomonas* (tequila)

Animal dung is known to contain a diverse range of bacteria, some of which might cause diseases in animals and people, while others are harmless. Foodborne pathogens such as *E. faecalis, P. vulgaris, P. aeruginosa, S. marcescens, S. typhi, S. epidermidis*, methicillin-resistant *S. aureus, B. anthracis, B. cereus*, and *S. pyogenes* continue to be a significant and recurring problem in the selected community due to a lack of knowledge. These pathogens can cause various infections, including hospital infections, endocarditis, bacteremia, urinary infections, genital duct infections, gastroenteritis, pneumonia, skin and urinary infections septicemia, respiratory and gastroenteritis infections, food poisoning, sore throat, rheumatoid fever, gangrene, scarlet fever, and acute glomerulonephritis.[4]

Many studies have been undertaken to understand the strategies for reducing bacterial development and its ramifications, highlighting the relevance of this information. Scientists have focused on using simple structured *Escherichia coli* (*E. coli*) bacterium as a model for understanding several fundamental biological processes in experimental trials. When comparing the cell envelopes of gram-negative and gram-positive bacteria, we will utilize *S. aureus* as a reference point. However, we will specifically focus on the differences between *S. aureus* and *B. subtilis*. *B. subtilis* is the primary gram-positive model organism and has a significant amount of knowledge associated with it. However, the cell envelope of *S. aureus* has been more widely investigated due to the interest in understanding how its surface features facilitate interactions with the environment during infection. It is important to exercise caution when making generalizations based on examples collected from specific bacteria. For instance, *E. coli* resides in the gastrointestinal tract of mammals. Hence, *E. coli* and other enteric bacteria require a cell membrane that is highly efficient in preventing the entry of detergents like bile salts. This may not be a significant concern for other gram-negative bacteria, as their cell envelopes may vary in species- and environment-specific manners. However, the fact that the Gram stain may be used to classify bacteria indicates that the fundamental organizing principles we provide are preserved. Furthermore, numerous bacteria possess an outermost layer known as the S-layer, which consists of a singular protein that completely envelops the cell. The evaluation does not cover S-layers and capsules, which are protective layers made of polysaccharides. Nanotechnology has the potential to be one of these alternate options. The implementation of potent novel antimicrobial drugs is crucial for managing pathogenic microorganisms, particularly antibiotic-resistant strains. This study specifically examined nanoparticles (NP) and nanofluids because of their growing prevalence in numerous commercial items and their potential usage as a relatively safe substance in diverse applications.[5]

7.5 ANTIMICROBIAL ACTIVITY OF NANOPARTICLES

Various types of nanoparticles have been examined for their ability to kill bacteria because of their distinct characteristics and methods of operation. Several nanoparticles that are frequently studied for their antibacterial properties include:

1. **Ag nanoparticles**: Silver nanoparticles have been intensively researched due to their powerful antibacterial capabilities. They have the ability to disturb the membranes of bacterial cells, hinder cellular respiration, and trigger oxidative stress by producing reactive oxygen species (ROS).

Important Note: **ROS** are very reactive molecules composed of oxygen atoms that are produced as inherent byproducts of biological metabolism. They have significant functions in diverse physiological processes, including cell signaling, immunological responses, and the control of gene expression. Nevertheless, an overabundance of ROS can result in oxidative stress, which can harm cellular constituents such as proteins, lipids, and DNA. The following are important categories of ROS:

- **Superoxide anion (O_2•-)**: Superoxide anion is produced by partially reducing oxygen during cellular respiration. It serves as a main precursor to other ROS and has the ability to directly harm cellular components or undergo conversion into other ROS, such as hydrogen peroxide.
- **Hydrogen peroxide (H_2O_2)** is produced through the dismutation of superoxide anion or by the action of enzymes like superoxide dismutase. It has the ability to pass through cellular membranes and function as a signaling molecule, or it can undergo additional metabolic processes to form more reactive species, such as hydroxyl radicals.
- **Hydroxyl radical (•OH)** is a highly reactive ROS that is formed through the Fenton reaction or Haber–Weiss reaction. These reactions involve the reaction of hydrogen peroxide with transition metal ions, such as iron or copper. Hydroxyl radicals have the potential to inflict significant harm on biomolecules such as DNA, proteins, and lipids.
- **Singlet oxygen ($^1O_2$)**: Singlet oxygen is produced when molecular oxygen interacts with photosensitizers or during the process of photosynthesis. It exhibits a high level of reactivity and has the potential to induce oxidative harm to cellular constituents.
- **Alkoxyl and peroxyl radicals (RO• and ROO•)** are produced when organic molecules are oxidized by other ROS. They have the ability to spread oxidative damage by extracting hydrogen atoms from nearby molecules, resulting in a series of chain reactions.
- **Hypochlorous Acid (HOCl)**: Hypochlorous acid is generated by neutrophils and other immune cells through the respiratory burst as a component of the innate immune response. It demonstrates antimicrobial activity by the process of oxidizing and causing damage to bacteria cells.

Cellular homeostasis and prevention of oxidative stress-related damage rely on maintaining a delicate equilibrium between the formation of ROS and the activity of antioxidant defense systems. An imbalance in levels of ROS has been linked to several clinical states, including aging, neurological illnesses, cardiovascular diseases, cancer, and inflammatory disorders. Gaining knowledge about the functions of ROS in maintaining good health and causing diseases is crucial for devising approaches to control ROS levels and reduce the harm caused by oxidative stress.[6]

2. **Copper and copper oxide nanoparticles:** Cu and CuO nanoparticles possess a wide range of antibacterial properties as they damage the membranes of bacterial cells, interfere with cellular enzymes and proteins, and induce oxidative stress. They exhibit a high level of efficacy specifically against gram-negative bacteria.

3. **Zinc oxide nanoparticles:** Zinc oxide nanoparticles exhibit antibacterial capabilities due to their capacity to generate ROS, break bacterial cell membranes, and hinder cellular respiration. They have efficacy against both gram-positive and gram-negative bacteria.
4. **Titanium dioxide nanoparticles:** TiO_2 nanoparticles have photocatalytic antibacterial properties when exposed to UV light. ROS, such as hydroxyl radicals, are produced. These radicals have the ability to harm bacterial cell membranes and biomolecules, resulting in the deactivation of bacteria.
5. **Gold nanoparticles:** Au nanoparticles have been investigated for their antibacterial properties; however, the exact mechanisms by which they work are not completely understood. They have the ability to engage with bacterial cell membranes, disturb cellular processes, and trigger oxidative stress.
6. **Carbon-based nanomaterials**, such as CNTs and graphene oxide (GO), have demonstrated antibacterial properties by causing physical harm to bacterial cell membranes, interfering with cellular activities, and producing ROS.
7. **Chitosan nanoparticles**, which are generated from chitin found in the exoskeletons of crustaceans, possess antibacterial properties as a result of their cationic composition. They have the ability to disturb the integrity of bacterial cell membranes, impede the proliferation of bacteria, and augment immunological responses.
8. **Polymeric nanoparticles:** These are manufactured from polymers such as polyethylene glycol, poly (lactic-co-glycolic acid), and polycaprolactone and have been designed to exhibit antibacterial characteristics. These nanoparticles have the ability to transport antibacterial substances, release compounds that kill microorganisms, or interfere with the membranes of bacterial cells.
9. **Metal oxide nanoparticles**, including magnesium oxide (MgO NPs), iron oxide (Fe_3O_4 NPs), and cerium oxide (CeO_2 NPs), have shown antibacterial properties by generating ROS and interfering with bacterial cell membranes. In addition to zinc oxide and titanium dioxide, these metal oxide nanoparticles exhibit antibacterial activity.

Nanoparticles have favorable prospects for creating innovative antimicrobial substances for diverse uses, such as medical instruments, bandages, antimicrobial layers, and water filtration systems. Nevertheless, it is imperative to conduct comprehensive assessments of their safety, efficacy, and potential environmental consequences before their broad use.

7.6 DISINFECTION MECHANISM OF WASTEWATER USING NANOFLUIDS (DISPERSION FORM OF NANOPARTICLES)

Nanoparticles in the dispersion form of nanofluids exhibit potent antibacterial properties against both gram-positive and gram-negative bacteria. Contrary to conventional disinfectants, these antimicrobial nanoparticles do not possess strong oxidizing properties. Therefore, it is anticipated that they will not generate detrimental DBPs. If utilized correctly in treatment procedures, they possess the capacity to substitute or augment conventional disinfection techniques. Here are some primary methods by which nanoparticles exhibit antibacterial effects:

- **Physical damage:** Nanoparticles have the ability to directly contact with microbial cells, causing disruption to their cell membranes or walls. This disruption can result in the release of cellular contents, compromised structural integrity, and eventually cell demise. Nanoparticles possess a small physical size and a large surface area, which greatly augment their capacity to infiltrate microbial cells.
- **Production of ROS:** Specific nanoparticles, such as silver (Ag), copper (Cu), zinc oxide (ZnO—**Figure 7.5**), and titanium dioxide (TiO_2), possess inherent characteristics that enable them to produce ROS when they come into contact with microbial cells. ROS, including superoxide radicals (O_2•-), hydroxyl radicals (•OH), and singlet oxygen ($^1O_2$), cause oxidative stress in microbial cells. This stress results in harm to proteins, lipids, and DNA, ultimately leading to the death of the cells.
- Metal nanoparticles, such as silver, copper, and zinc nanoparticles, have the ability to release metal ions into their surrounding environment. The presence of these metal ions can interfere with crucial biological functions, such as the functioning of enzymes and the replication of DNA, ultimately resulting in the death of microbial cells. The liberation of metal ions is frequently contingent on pH and can be intensified in acidic circumstances commonly encountered in microbial habitats.
- **Surface contact killing:** Nanoparticles, such as silver nanoparticles, that have a large surface-to-volume ratio, can directly interact with microbial

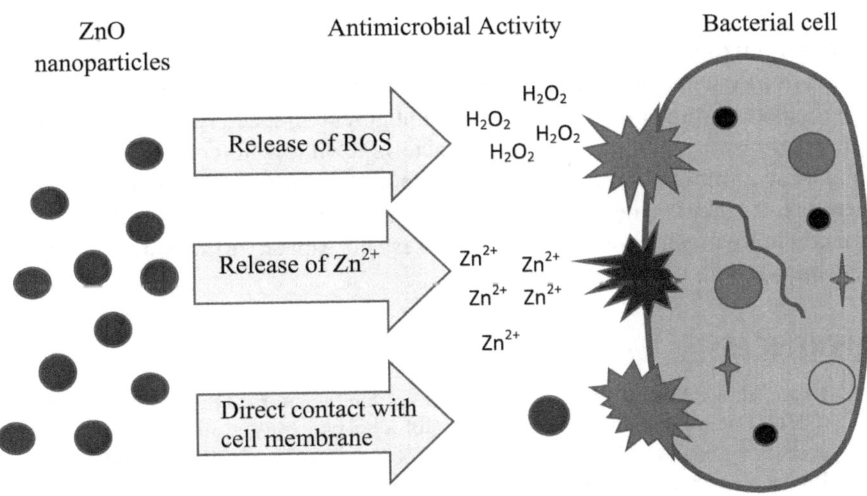

FIGURE 7.5 Disinfection mechanism of ZnO nanofluids.[7]

cells when they come into contact, causing them to become inactive. The surface contact killing mechanism is highly efficient in eliminating bacteria and fungi that directly interact with surfaces coated with nanoparticles.

- Nanoparticles have the ability to imitate the antibacterial properties of endogenous peptides by interfering with the cell membranes of microorganisms. Nanoparticles, often known as nano-antimicrobials, can be designed to selectively attack certain microbial infections while limiting harm to host cells.
- Nanoparticles can have synergistic antimicrobial effects when used in conjunction with conventional antimicrobial agents, such as antibiotics or antiseptics. Combining nanoparticles with other antimicrobial treatments can increase their effectiveness and decrease the emergence of antimicrobial resistance.

7.7 CHALLENGES

Several parameters, such as nanoparticle size, shape, surface chemistry, concentration, and the specific microbial species being targeted, might influence the antibacterial activity of nanoparticles. Before widespread implementation in antimicrobial applications, it is crucial to thoroughly assess the safety, effectiveness, and potential ecological consequences of nanoparticles, notwithstanding their significant potential as antimicrobial agents. Furthermore, continuous research is being conducted to enhance the effectiveness of antimicrobial compositions that include nanoparticles and to devise methods to minimize any potential dangers related to their utilization.

7.8 SUMMARY

Several factors, such as the type and amount of pollution in the wastewater, legal obligations, practical considerations, and environmental concerns, all contribute to the selection of the appropriate disinfection technique. Wastewater treatment plants commonly use multiple disinfection procedures to effectively eliminate pathogens while minimizing environmental impact. Regular monitoring and maintenance of disinfection systems are crucial to ensure their effective disinfection and compliance with regulatory standards. Nanofluids exhibit potent antibacterial properties against both gram-positive and gram-negative bacteria. Contrary to conventional disinfectants, these antimicrobial nanoparticles do not possess strong oxidizing properties. Therefore, it is anticipated that they will not generate detrimental DBPs. If utilized correctly in treatment procedures, they possess the capacity to substitute or augment conventional disinfection techniques.

REFERENCES

1. Khan, M. T., Ahmad, R., Liu, G., Zhang, L., Santagata, R., Lega, M., & Casazza, M. (2024). Potential environmental impacts of a hospital wastewater treatment plant in a developing country. *Sustainability, 16*, 2233. https://doi.org/10.3390/su16062233
2. Jiménez-Jiménez, C., Moreno, V. M., & Vallet-Regí, M. (2022). Bacteria-assisted transport of nanomaterials to improve drug delivery in cancer therapy. *Nanomaterials, 12*(2), 288. https://doi.org/10.3390/nano12020288

3. Sartini, S., Permana, A. D., Mitra, S., Tareq, A. M., Salim, E., Ahmad, I., Harapan, H., Bin Emran, T., & Nainu, F. (2021). Current state and promising opportunities on pharmaceutical approaches in the treatment of polymicrobial diseases. *Pathogens*, *10*(2), 245. https://doi.org/10.3390/pathogens10020245
4. Malika, M., Rao, C. V., Das, R. K., Giri, A. S., & Golder, A. K. (2016). Evaluation of bimetal doped TiO2 in dye fragmentation and its comparison to mono-metal doped and bare catalysts. *Applied Surface Science*, *368*, 316–324.
5. Gujar, J. G., Patil, S. S., & Sonawane, S. S. (2023). A review on nanofluids: Synthesis, stability, and uses in the manufacturing industry. *Current Nanomaterials*, *8*(4), 303–318.
6. Thakur, P. P., Sonawane, S. S., & Mohammed, H. A. (2023). Recent trends in applications of nanofluids for effective utilization of solar energy. *Current Nanoscience*, *19*(2), 170–185.
7. Dimapilis, E. A. S., Hsu, C.-S., Mendoza, R. M. O., & Lu, M.-C. (2018). Zinc oxide nanoparticles for water disinfection. *Sustainable Environment Research*, *28*(2), 47–56. https://doi.org/10.1016/j.serj.2017.10.001.

8 The Application of Hybrid Nanofluids in the Field of Membrane Technology

Due to the growing demand for rare earth metals, there has been a significant surge in interest in the process of separating and purifying these compounds. Various methods, such as chemical precipitation, ion exchange, and solvent extraction, can be used to recover these metals from water-based solutions. The conventional liquid–liquid extraction (LLE) technique is considered to be time-consuming and necessitates a substantial quantity of costly organic solvents to achieve the desired outcomes. Membrane-based technology is a cost-effective and efficient substitute for solvent extraction, and it plays a crucial role in the separation and purification processes conducted across several industries.

8.1 CLASSIFICATION OF LIQUID MEMBRANES

Membrane operations are regarded as environmentally sustainable and are endorsed as the most effective techniques. This is because they are straightforward and efficient in both design and operation, can be easily expanded, and have low energy consumption. They also have a significant potential for positive environmental and energetic effects. A membrane is a selective barrier that allows the movement of solutes between two phases. Liquid membranes (LMs) are a type of membrane that consists of a liquid that is not soluble in the feed (donor phase) and receiving (acceptor) phases. This liquid acts as a semipermeable barrier between the two phases, whether they be liquid or gas. LMs are categorized based on their module configuration, transport mechanism, application, type of carrier, and kind of membrane support.

Various carrier fluids commonly used in LLE are employed as commercial extractants. Most importantly, the carrier fluid must demonstrate selectivity toward the desired solute, as well as possessing properties of being non-flammable, non-toxic, biodegradable, cost-efficient, stable, soluble in the organic phase, poorly soluble in water, and easily regenerable. Typically, there are three primary categories of carriers: acidic, basic, and neutral. The subsequent are the primary configurations of carrier fluids for LMs.

- **Acidic carrier:**
  The transportation process relies on the interaction between a metal cation and an acidic extractant, resulting in the formation of a neutral complex. In this case, hydroxy oximes and alkyl phosphoric acids are primarily

DOI: 10.1201/9781003595137-8

employed. HDEHP or DEHPA (di(2-ethylhexyl) phosphoric acid) and di-(2-ethyl hexyl) phosphoric acid (D2EHPA) are some of the examples of acidic carrier fluids, which are extensively studied in the extraction of Cu (II), Co (II), Cd (II), Ni (II), Zn (II), cobalt (II), europium (III), yttrium (III), and neodymium (III).

- **Basic carrier:**
  The transportation primarily relies on the process of addition or ion exchange reactions. This involves the formation of anion pairs, with common carriers being tri-octyl amine and tri-decyl amine carriers are extensively utilized carriers for the extraction of Cd (II), U (VI), Pt (IV), and Pd (II) heavy metals.
- **Neutral carrier:**
  The transportation process relies on solvation, resulting in the formation of a solvate or adduct. These compounds consist of trialkyl phosphine oxides or phosphoric acid esters/tri-octyl phosphine oxide, tri-butyl phosphate, and tetra-n-octyl diglycol amides are often utilized neutral carriers.

According to reports, a solitary entity has the capability to integrate extraction and stripping operations by employing LMs. These membranes offer various advantages, including non-equilibrium mass transfer, exceptional selectivity, high flux rates, and low energy consumption. There are generally three categories of liquid membranes, which are BLM, SLM, and ELM. BLM is an acronym for bulk liquid membranes, SLM stands for supported liquid membranes, and ELM stands for emulsion liquid membranes. The following are the main configurations of LMs.

8.2 SUPPORTED LM EXTRACTION

SLM extraction involves the use of a thin membrane to separate the aqueous feed and stripping phases as shown in **Figure 8.1**. The membrane contains an immobilized organic phase, and the transport of metal species is facilitated by organic

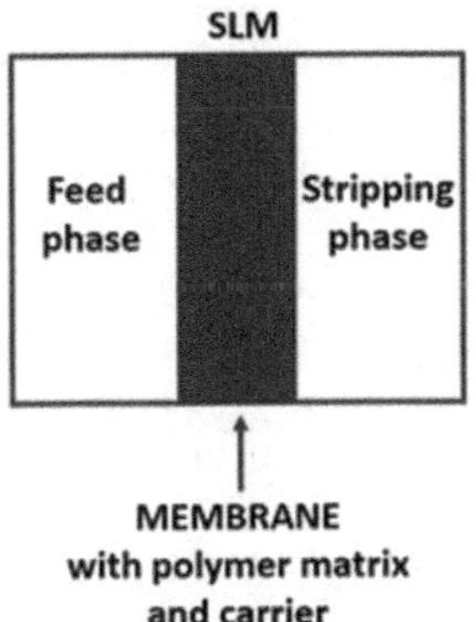

Application of:

- well-known carriers (e.g., TOPO, D_2EHPA) when changing the process conditions (e.g., composition of feed and stripping phases),
- new carriers (e.g., lipophilic phosphorylated betaine),
- various ionic liquids (e.g., $[C_6MIM][NTf_2]$) as carriers,
- synergistic effect of a combination of ionic liquids and traditional extractants as carriers (e.g., ionic liquid $[C_6MIM][NTf_2]$ and the mixture of TOPO and TBP),
- advanced mathematical models for analyzing REE ions separation to determine optimal experimental conditions.

FIGURE 8.1 Schematic diagram of typical SLM along with its applications.[1]

carriers present in the micropores of the support. The utilization of solvent extraction across an SLM can be conceptualized as a concurrent three-step procedure involving the extraction of metal ions from the initial phase to the SLM, the diffusion of ions across the SLM, and the subsequent stripping process to the receiving phase. Selective transport in this context refers to the movement of specific ions through a material. This movement is influenced by the ion permeability and thickness of the SLMs. It is important to acknowledge that the choice of membrane form can also impact the membrane procedure. The properties of the carriers utilized are a significant component that influences the SLM processes. Consequently, numerous research has been done to explore the feasibility of employing different chemical compounds for this objective. At present, established carriers are being utilized, with alterations made to the parameters of membrane processes.

However, there is also a search for new chemicals that may offer improved efficiency as carriers. Carriers are typically categorized into classes based on their distinct features, including basic, acidic, chelating, macrocyclic, macromolecular, neutral, and solvating chemicals. D2EHPA is a commonly utilized acidic carrier in several LMs designed for the extraction of metal ions. D2EHPA has several benefits, including a strong preference for rare earth metals and compatibility with a wide range of common diluents. Recently, D2EHPA has been employed in multiple techniques aimed at extracting diverse heavy metals from various sources.

8.3 EMULSION LM EXTRACTION

In recent years, researchers have sought alternative LMs due to technological issues with the stability of supported LMs. ELM techniques have significant promise for extracting and separating metal ions across a wide range of concentrations, from extremely low to high. This is due to the fact that, when compared to LLE, they necessitate a reduced volume ratio of the organic phase to the aqueous feed solution. In addition, they facilitate a very rapid transfer of mass because of the extensive surface area present in the emulsion globules and interior droplets. ELMs are created by emulsifying two phases that do not mix, such as an oil-in-water-in-oil (O/W/O) system or a water-in-oil-in-water (W/O/W) system. **Figure 8.2** depicts these two main classifications of multiple emulsions that are now accessible. The fundamental driving force for solute flow through the membrane is the disparity in **chemical potentials** between the external feed phase and the internal phase, similar to other LMs.

An ELM is a specific LLE technology designed for the separation and purification of compounds from aqueous solutions. The process is comprised of three stages (**Figure 8.3**): the **feed phase** (which contains an aqueous solution with the desired solutes), the **organic phase** (which is an immiscible solvent), and the **stripping phase** (which is an aqueous solution used to extract the solutes).

An ELM system is constructed by distributing minute droplets of the organic phase throughout the feed phase, resulting in the formation of a stable emulsion. The solutes of interest migrate from the feed phase to the scattered droplets of the organic phase through either diffusion or chemical reactions. The solutes are subsequently

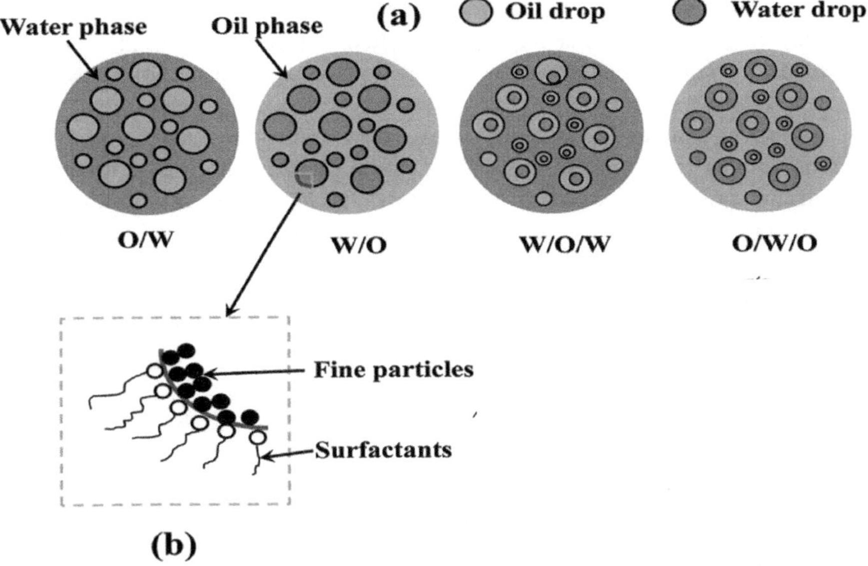

FIGURE 8.2 Schematic diagram of (a) water-in-oil-in-water (W/O/W) and oil-in-water-in-oil (O/W/O) emulsions and (b) oil–water emulsion interface.[2]

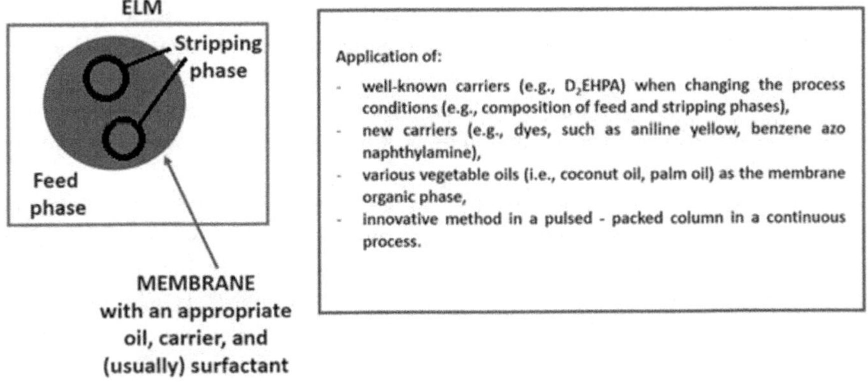

FIGURE 8.3 Schematic diagram of typical emulsion liquid membrane (ELM) along with its applications.[1]

sent through the organic phase to the stripping phase, where they are specifically removed. Essential elements and characteristics of ELMs encompass:

- **Organic phase:** The organic phase refers to a solvent that is not soluble in water and contains extractants or carriers that specifically bond to the desired solutes. Typical organic solvents utilized in ELM systems comprise kerosene, aliphatic hydrocarbons, and fluorinated chemicals.

- **External phase:** The external phase refers to the aqueous solution that contains the solutes to be separated or purified. The scattered droplets of the organic phase are encompassed by it, and it acts as the origin of solutes for extraction.
- **Stripping phase:** The stripping phase refers to an aqueous solution where the solutes are removed from the organic phase. Commonly, favorable circumstances are selected to facilitate the separation of solutes from carriers or extractants, such as adjusting the pH or adding complexing agents.
- **Surfactants:** Surfactants are incorporated into the system to enhance the stability of the emulsion and inhibit the merging of the scattered droplets. They provide a defensive barrier surrounding the droplets, decreasing the tension between their surfaces and improving the stability of the emulsion. Surfactants might additionally impact the mass transport characteristics of the system.
- **Carriers or extractants:** They are chemical substances that are introduced into the organic phase to aid in the specific extraction of desired solutes. These chemicals adhere to the desired solutes and facilitate their transfer through the organic phase.

8.4 APPLICATIONS OF ELM

ELMs have various applications, such as metal ion extraction. ELM systems are employed for the retrieval and refinement of metal ions from aqueous solutions, such as wastewater generated by mining activities or industrial processes. ELM systems are capable of extracting organic substances, such as phenols, organic acids, or dyes, from aqueous solutions. This extraction method is used for environmental remediation or chemical processing purposes.

ELM systems have been utilized to segregate and refine biological molecules, including proteins, enzymes, and medicines, from intricate aqueous mixtures. ELMs provide a flexible and effective method for separating and purifying chemicals from water-based solutions. They have wide-ranging uses in industries such as environmental protection, chemical manufacture, and biotechnology. Some of the applications include:

- Heavy metal or rare earth metal separation or concentration
- Biomaterial, organic compound, and pharmaceutical product recovery and separation
- Wastewater treatment
- Extraction of various metal ions, such as cadmium (II), cobalt (II), manganese (II), nickel (II), chromium (III), chromium (IV), cerium (IV), mercury (II), uranium (VI), silver (I), palladium (II), and gallium (III)
- Removal of numerous types of metal ions: Cu(II), Zn(II), Cd(II), Co(II), Mn(II), Ni(II), Cr(III) and Cr(IV), Ce(IV), Hg(II), U(VI), Ag(I), Pd(II), Ga(III)

8.5 ADVANTAGES OF ELM

ELMs provide numerous benefits in diverse separation and purification procedures:

i. **Selective extraction:** ELMs have the ability to extract specific solutes, such as metal ions or organic molecules, from intricate mixtures, while leaving other components behind. The selectivity is attained by carefully selecting suitable extractants and carrier phases inside the LM.
ii. **Enhanced efficiency:** ELMs can attain remarkable extraction efficiencies as a result of the substantial interfacial area between the membrane phase and external feed phase. This enables the efficient movement of solutes across the membrane.
iii. **Flexibility:** ELMs can be customized to separate various solutes, such as metal ions, organic contaminants, and even gases. Their adaptability makes them well-suited for a wide range of uses in the fields of environmental cleanup, chemical manufacturing, and the extraction of valuable resources.
iv. **Minimized environmental impact:** In comparison to conventional methods like solvent extraction or distillation, ELMs can significantly decrease the use of solvents and energy, resulting in a decreased environmental footprint.
v. **Operational simplicity:** ELM processes are characterized by their straightforward operation and seamless integration into pre-existing industrial processes. Standard equipment can be used to create and sustain emulsions, and the separation of the different phases can be accomplished by straightforward methods such as gravity settling or centrifugation.
vi. **Continuous operation:** ELM operations can be conducted without interruption, enabling the continuous extraction and separation of solutes from the feed solution. This can result in increased throughput and productivity compared to batch processes.
vii. **Scalability:** ELM methods may be quickly expanded from small-scale laboratory settings to large-scale industrial settings, making them well-suited for industrial applications and large-scale production.
viii. **Reusability:** The components of the LM, such as extractants or carriers, can be regenerated and reused in certain situations. This results in cost savings and a decrease in waste production.

In general, ELMs present a hopeful method for addressing different separation and purification difficulties, as they combine excellent effectiveness with environmental friendliness and ease of use.

8.6 LIMITATIONS OF ELM

ELMs have numerous benefits for different separation and purification procedures. However, they also possess certain drawbacks:

i. **Emulsion stability:** The stability of emulsions can be difficult to maintain, particularly when working in turbulent settings or when there are surfactants or contaminants present. Emulsions have the potential to combine and form larger droplets, resulting in the separation of different phases and a decrease in effectiveness.

ii. **Mass transfer limitations**: The movement of solutes across the LMs might be hindered, especially for big molecules or when the liquid is highly viscous. This can impede the overall throughput and efficiency of the process.

iii. **Membrane fouling:** It refers to the accumulation of impurities or solid particles on the LM interface. This can result in decreased extraction effectiveness and higher maintenance needs.

iv. **Selectivity challenges:** Attaining a high level of selectivity for desired solutes can be difficult, particularly in situations involving complex mixtures where other substances may vie for extraction. Precision is essential in applications such as metal ion extraction.

v. **Environmental concerns:** Improper management or disposal of some chemicals used in ELM systems, such as organic solvents or extractants, might lead to environmental concerns.

vi. **Challenges in scaling up:** The process of scaling up ELM technologies from a small laboratory size to a larger industrial scale can be intricate and expensive due to the engineering difficulties related to maintaining emulsion stability, facilitating mass transfer, and optimizing the overall process.

vii. **Energy requirements:** ELM methods may necessitate substantial energy inputs, especially for the production and separation of emulsions, which might influence the overall cost-effectiveness and environmental impact of the process.

The only limitation that needs to be addressed to resolve all other limitations related to ELMs is maintaining the stability of the emulsion. To address the problem of stability, researchers introduced nanoparticles into the surfactants. According to reports, the nanoparticles in the dispersion form can create a protective shell around the emulsion droplets, preventing them from merging together. This is how the origin of the emulsion nanofluid membrane (ENM) occurred.

8.7 INTRODUCTION TO ENM

The prominent drawbacks of ELM are the instability of the emulsion and the challenge of breaking it during the extraction process. The stability of ELM can be defined as the ability of an LM to withstand rupture when exposed to intense shear stress. Globule rupture can be caused by several processes, such as osmotic swelling, flocculation, Ostwald ripening, creaming, and droplet coalescence. To address these drawbacks, a novel ENM was created as an enhanced version of the traditional ELM method to eliminate harmful pollutants from water-based solutions. When comparing solid particles, which have a nanoscale size, to emulsion droplets, the solid particles are significantly smaller. This enables them to stable droplets of various sizes, ranging from small to large. The droplets are enveloped by a layer of nanoparticles, which exhibit substantial adsorption at the interface between the oil and water. Moreover, this layer offers a steric barrier to prevent the merging of droplets, hence contributing to their stability. The arrangement of particles in nanoparticle-stabilized emulsions is governed by the minimum free energy, which prevents the particles from sinking when they are trapped at the boundary between different liquids.

8.7.1 Procedure for the Preparation of ENM

The process for developing ENMs consists of multiple stages and may differ based on the intended use and desired characteristics of the membrane ENM. As shown in **Figure 8.4**.

- The external phase consists of the wastewater from which the heavy metal or solute is to be removed.
- The membrane phase consists of a mixture of surfactant, carrier fluid, nanofluid, and extractant.
- The stripping phase consists of the stripping agent. It is alternatively referred to as the internal phase.

Further, a comprehensive outline of the procedure is given here:

i. **Component selection:** Identify the components of the ENM according to the desired application. Typically, this involves choosing an organic phase as a carrier fluid, an extractant or functionalizing agent, and nanoparticles to improve stability or provide added functionality.

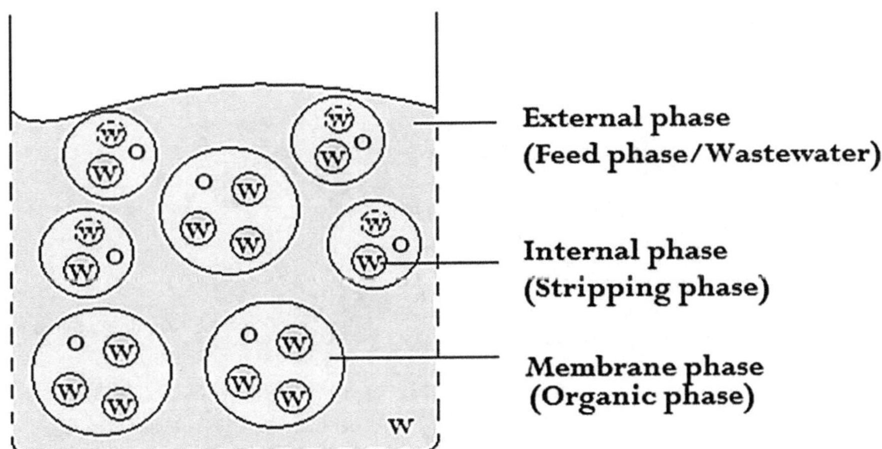

FIGURE 8.4 Understanding the process of water-in-oil-in-water (W/O/W) emulsion formation.[3]

ii. **Nanoparticle dispersion preparation:** Introduce nanoparticles (such as metal oxides, CNTs, or graphene oxide) into the carrier fluid to create a homogeneous mixture. These objectives can be accomplished via techniques such as ultrasonication, mechanical stirring, or sonication-assisted dispersion.[4]

iii. **Emulsification (W/O):** Combine nanofluid (the nanoparticle dispersion) with the aqueous phase to create a stable emulsion as shown in **step 1** of **Figure 8.5**. These methods include mechanical agitation, high-pressure homogenization, or sonication. To ensure a consistent dispersion of nanoparticles within the emulsion droplets, it is necessary to optimize the emulsification process.

iv. **LM formation (W/O/W):** Induce the emulsion to undergo phase separation, resulting in two distinct phases: the dispersed phase containing the nanoparticles and the continuous phase (aqueous phase). The droplets in the emulsion include nanoparticles and together they create an LM as shown in **step 2** of **Figure 8.5**.

v. **Stabilization:** Enhance the stability of the emulsion and prevent the merging of droplets by using surfactants or stabilizers. The selection of surfactant is contingent upon the characteristics of both the carrier fluid and the nanoparticles employed.[5]

vi. **Characterization:** Evaluate the stability, particle distribution, and rheological features of the ENM through characterization. Characterization can be accomplished by the utilization of techniques such as dynamic light scattering, scanning electron microscopy, and rheological analysis. The droplet size of the emulsion was analyzed using an optoscope microscope. After preparing the samples, a small amount was placed on a glass slide and viewed under a microscope. Images were acquired at different magnifications using the Celeste microscopy program (version 3.2). As shown in **Figure 8.6**, the addition of nanoparticles is surrounded by the emulsion droplet to enhance the stability.

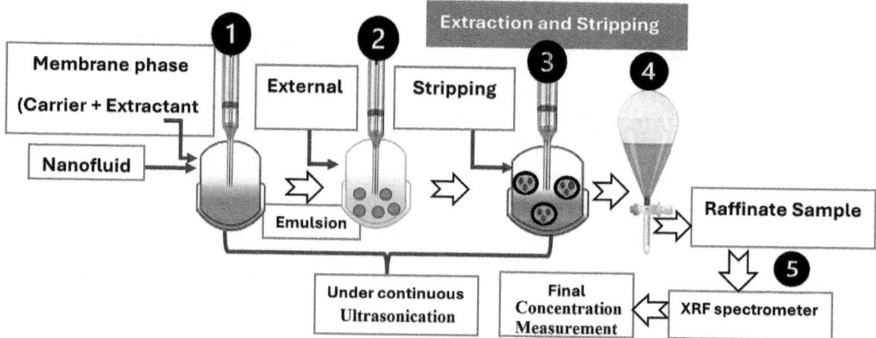

FIGURE 8.5 Procedure for the extraction of pollutants using emulsion nanofluid membrane (ENM).

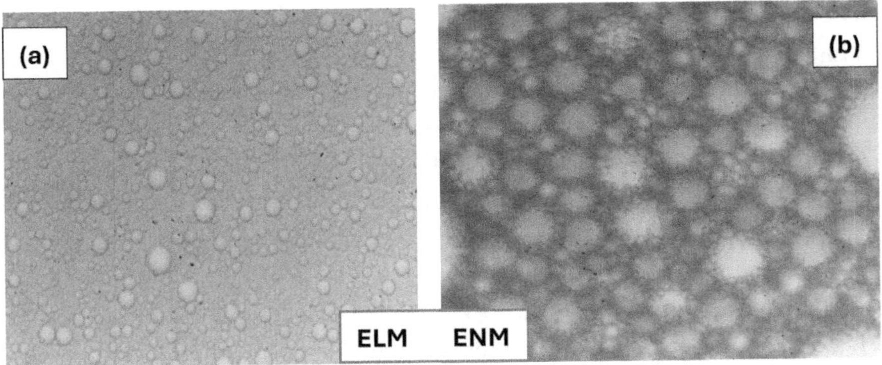

FIGURE 8.6 Optical spectroscopy results of emulsion membrane: (a) emulsion liquid membrane (ELM) and (b) emulsion nanofluid membrane (ENM).

vii. **Implementation:** Utilize the ENM in the specific separation or purification procedure. One such method is to submerge a permeable support material into the emulsion to form an SLM. Alternatively, the emulsion itself can be used as a LM for extraction or filtration procedures.

viii. **Optimization and Testing:** Enhance the characteristics of the ENM by conducting experimental tests and optimizing the parameters. To achieve the desired performance, it may be necessary to modify the nanoparticle concentration, surfactant type and concentration, or emulsification conditions. The emulsion that had been formulated was evenly distributed into the mixing chamber along with the feed phase solution and agitated for approximately ten minutes as shown in **step 3** of **Figure 8.5**. After the completion of the mixing process, the solution was left to undergo gravitational separation in a separating funnel (**step 4** of **Figure 8.5**). Raffinate samples of approximately 5 ml were extracted from the solution, and the concentration of the pollutant was analyzed every 5 minutes of contact time using a UV spectrometer and X-ray fluorescence (XRF) analysis (**step 5** of **Figure 8.5**). The extraction percentage can be determined using **Equation 8.1**.

$$\text{Extraction } \% = \frac{\left(\text{Initial concentration} - \text{Final Concentration}\right)_{\text{Pollutant}}}{\text{Initial concentration of pollutant}} \tag{8.1}$$

The proportion of membrane breaking in the emulsion was determined by analyzing the concentration of H^+ ions in the external feed solution, taking into account the volume of the internal phase, using specific **Equations 8.2 and 8.3**.

$$\text{Membrane breakage}(\%) = \frac{V_s}{V_i} \times 100 \tag{8.2}$$

where

V_i : Initial volume of the internal phase
V_s : Volume of the internal phase leaked into the external phase

$$V_s = V_{Ext} \frac{10^{-pH_i} - 10^{-pH_F}}{10^{-pH} - C_{OH^-}{}^{i}} \tag{8.3}$$

where

V_{Ext} : Initial volume of the external phase
pH_i : Initial pH of the external phase
pH_F : pH of the the external phase after addition of emulsion after stirring
$C_{OH^-}{}^{i}$: Initial concentration of OH^- in the internal phase

ix. **Scale-up:** Increase the production of the ENM for industrial applications, while maintaining a consistent and reproducible manufacturing process.

By adhering to these procedures and optimizing the composition and preparation circumstances, ENMs can be customized for particular uses such as wastewater treatment, gas separation, or extraction processes. This customization results in improved stability and functionality when compared to conventional LMs.

8.7.2 Extraction Mechanism of ENM

The extraction process of an ENM involves the specific transfer of solutes (such as ions or organic compounds) from the initial phase to the stripping phase through the LM phase, which comprises nanoparticles scattered inside the emulsion droplets. Nanoparticles are essential for improving the efficiency and selectivity of the membrane during extraction. Here is a summary of the extraction mechanism (**Figure 8.7**):

- **Partitioning:** The act of dividing something into separate parts or sections. When the feed solution, which contains the solutes, interacts with the ENM, the solutes distribute themselves between the feed phase and

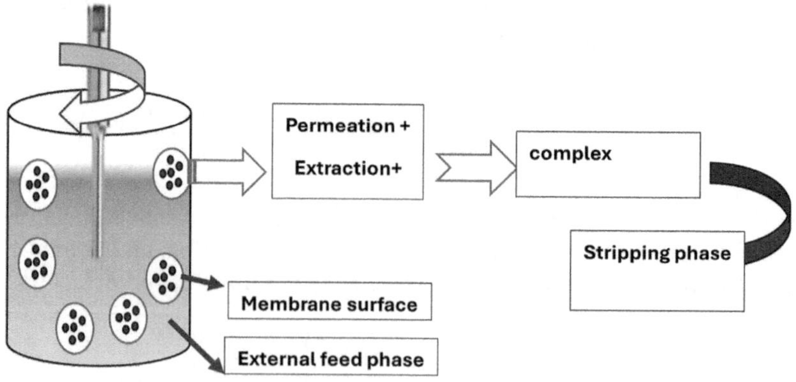

FIGURE 8.7 Extraction mechanism of emulsion nanofluid membrane (ENM).

the LM phase. The solutes permeate the LM phase through diffusion, driven by their partition coefficients, which are influenced by parameters such as the solute's affinity for the carrier fluid and the surface chemistry of the nanoparticles.[6]

- **Diffusion:** Solutes undergo diffusion via the nanoparticle-rich environment within the LM phase. Nanoparticles in an LM enhance the tortuosity of the diffusion path, resulting in a slower diffusion rate compared to a membrane without nanoparticles. The large surface area of nanoparticles enables efficient interaction with solutes, enhancing their transport across the membrane.
- **Chemical interaction:** Chemical interactions may take place at the surface of nanoparticles, depending on the characteristics of the nanoparticles and solutes involved. This may entail processes such as complexation, ion exchange, or other chemical reactions that improve the efficiency and selectivity of the membrane in extracting substances.
- **Interfacial interaction during the stripping phase:** As solutes diffuse through the LM phase, they ultimately reach the interface with the stripping phase, which is usually an aqueous solution. In this process, solutes migrate from the LM phase to the stripping phase as a result of disparities in concentration or affinity. This stage finalizes the extraction process by moving solutes from the feed phase to the stripping phase.
- **Desorption and regeneration:** In certain cases, it may be necessary to recover or regenerate the solutes that have been removed into the stripping phase. These objectives can be accomplished by employing techniques such as pH manipulation, solvent separation, or chemical formation, which are selected based on the characteristics of the substances involved and the intended result.

The extraction mechanism of an ENM utilizes principles of partitioning, diffusion, and chemical interaction to ensure effective and specific transport of solutes across the LM phase. Nanoparticles augment the membrane's effectiveness by amplifying its surface area, furnishing reactive sites, and enhancing stability.[7]

8.8 SUMMARY

Nanofluids are essential in ENMs, as they offer numerous advantages that improve the membrane's performance and functionality. The following are a few crucial factors regarding nanofluids in engineered nanomaterials (ENMs):

1. **Improved stability**: The presence of nanoparticles in the carrier fluid enhances the stability of the emulsion by creating a protective layer around the droplets. This inhibits the merging and growth of the emulsion droplets, resulting in enhanced stability of the LM phase.
2. **Enhanced interfacial area**: Nanoparticles contribute to an increased interfacial area between the membrane phase and external feed phase (wastewater) of the emulsion. This allows for a greater amount of surface area for the transfer of mass and contact with solutes, resulting in improved efficiency of extraction.

3. **Enhanced selectivity**: Nanoparticles have the ability to be modified with certain ligands or surface coatings that give the LM selective capabilities. This enables the specific removal of desired substances while preventing the presence of unwanted substances, hence enhancing the membrane's selectivity.
4. **Enhanced mass transfer via facilitation**: Nanoparticles have the ability to modify the flow characteristics of the LM, resulting in a decrease in viscosity and an increase in the rate at which mass is transferred. Consequently, this leads to enhanced solute diffusion across the membrane, resulting in increased extraction efficiencies and reduced processing durations.
5. **Customizable characteristics**: The characteristics of nanofluids, such as the size, shape, and surface chemistry of the particles, can be adjusted to match the specific needs of certain applications. This enables precise adjustment of the membrane's performance and selectivity for various solutes and operating circumstances.
6. **Improved chemical stability**: Nanoparticles can enhance the chemical stability of the LM by scavenging reactive species or avoiding degradation of the carrier fluid. This enhances the durability and strength of the membrane, particularly in challenging chemical conditions.
7. **Flexibility**: Nanofluids possess the ability to be utilized with a diverse range of carrier fluids and nanoparticles, allowing for adaptability in membrane design and optimization for various applications. This adaptability allows for the creation of customized membranes to address specific separation and purification difficulties.

Incorporating nanofluids into ENMs has several benefits, such as increased stability, higher rates of mass transfer, the capacity to selectively extract substances, and adjustable features. The advantages of ENMs make them highly suitable for a wide range of applications in wastewater treatment, chemical processing, environmental remediation, and resource recovery.

REFERENCES

1. Kaczorowska, M. A. (2023). The latest achievements of liquid membranes for rare earth elements recovery from aqueous solutions—A mini review. *Membranes*, *13*(10), 839. https://doi.org/10.3390/membranes13100839
2. Tian, Y., Zhou, J., He, C., He, L., Li, X., & Sui, H. (2022). The formation, stabilization and separation of oil–water emulsions: A review. *Processes*, *10*(4), 738. https://doi.org/10.3390/pr10040738
3. Björkegren, S., Karimi, R. F., Martinelli, A., Jayakumar, N. S., & Hashim, M. A. (2015). A new emulsion liquid membrane based on a palm oil for the extraction of heavy metals. *Membranes*, *5*(2), 168–179. https://doi.org/10.3390/membranes5020168
4. Khedkar, R. S., Sonawane, S. S., & Wasewar, K. L. (2012). Influence of CuO nanoparticles in enhancing the thermal conductivity of water and monoethylene glycol based nanofluids. *International Communications in Heat and Mass Transfer*, *39*(5), 665–669.
5. Khedkar, R. S., Sonawane, S. S., & Wasewar, K. L. (2013). Synthesis of TiO2–water nanofluids for its viscosity and dispersion stability study. *Journal of Nano Research*, *24*, 26–33.

6. Khedkar, R. S., Kiran, A. S., Sonawane, S. S., Wasewar, K. L., & Umare, S. S. (2013). Thermo-physical properties measurement of water based Fe3O4 nanofluids. *Carbon—Science and Technology*, *5*(1), 187–191.
7. Sonawane, S. S., Khedkar, R. S., & Wasewar, K. L. (2013). Study on concentric tube heat exchanger heat transfer performance using Al_2O_3–water based nanofluids. *International Communications in Heat and Mass Transfer*, *49*, 60–68.

9 The Application of Hybrid Nanofluids in the Field of Petroleum Science and Technology

Petroleum, also known as crude oil, is an essential natural resource that plays a crucial role in supporting the worldwide energy economy. Crude oil serves as the main fuel for transportation, acts as a fundamental raw material for the petrochemical sector, and plays a vital position in energizing different facets of contemporary existence. Petroleum extraction is a sophisticated and comprehensive process that involves extracting oil from subsurface reserves using a number of intricate procedures. It demands a thorough knowledge of geology, engineering, and technology. This chapter will examine the utilization of nanofluid techniques and procedures employed in the extraction of petroleum, investigating the essential stages and advanced technology utilized in this vital business.

Oil is commonly located in geological formations known as oil reservoirs, which are present in diverse places worldwide. The classification of oil reservoirs is based on their depth, which is categorized as follows: shallow (30–800 m), medium (800–2000 m), deep (2000–5000 m), and ultra-deep (greater than 5000 m). The classification is continuously evolving due to advancements in drilling technology that provide the opportunity to reach higher depths. Regardless of the oil reservoir's depth, the fundamental principle of oil extraction remains consistent and is rooted in the life cycle of the oil field (**Figure 9.1**).

The process of petroleum oil and gas extraction has five distinct stages: exploration, appraisal, development, production, and abandonment.

Exploration is a systematic approach employed to seek potentially profitable oil and gas reserves by conducting geological surveys and drilling exploration wells to identify areas with promising prospects. Throughout the drilling procedure, data and specimens are gathered to gain knowledge about the composition of the rocks and fluids. This information helps determine the potential quantity of oil and gas present in the surveyed region, as well as the depth at which the oil and gas reservoirs are located.

Following the successful completion of drilling exploration wells, the **appraisal stage** of the lifecycle commences. The primary objective of this phase is to enhance the field description by gathering additional data and minimizing the uncertainty or potential losses related to the size, shape, and marketability of the oil and gas reservoir.

The **development phase** follows a satisfactory appraisal and precedes manufacturing. The primary tasks involve creating a conceptual development plan to facilitate the growth of the oil and gas field. This includes designing production wells,

DOI: 10.1201/9781003595137-9

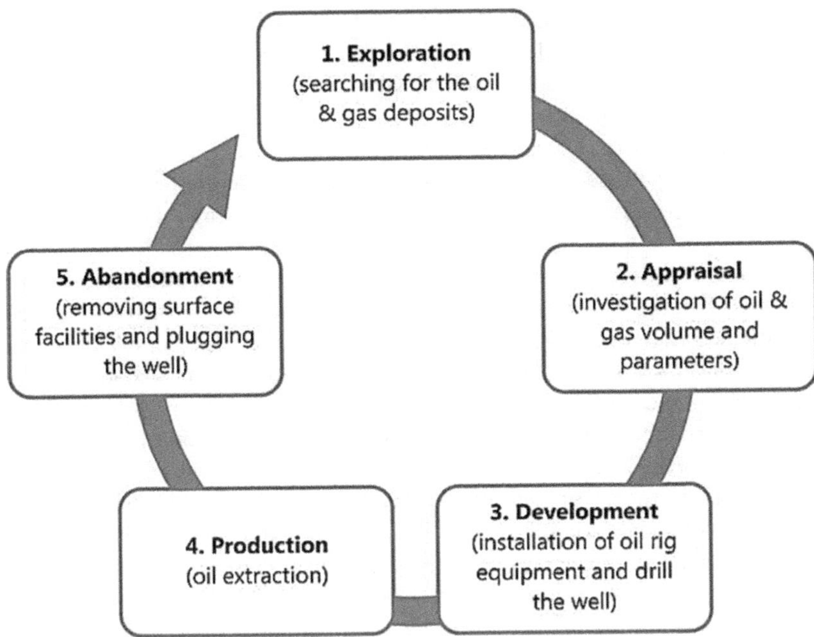

FIGURE 9.1 Diagram illustrating the sequential process of extracting crude oil.[1]

determining the necessary surface and subsurface facilities, and outlining the principles for operation and maintenance. Additionally, the construction of the facilities and production units is carried out.

The **production phase** commences with the initiation of the initial oil flow in the wellhead. The lifespan of oil and gas fields typically spans from 15 to 30 years, starting with the initial extraction of oil until the point of abandonment. However, in the case of the greatest reserves, this lifespan can be prolonged to 50 years or even longer. Once extracted, oil and gas are transported for processing and distribution.

Once oil and gas production becomes economically unviable, wells are sealed and **abandoned**, and production facilities are dismantled, marking the final phase of the life cycle of oil and gas fields.

The use of nanotechnology in oil extraction research has concentrated on solving the limits of present technology through the implementation of nano-assisted approaches. Nanomaterials have been widely employed in numerous industries. In the oil and gas business, these nanoparticles are generally used as nanofluids, which are liquid suspensions containing diluted nanoparticles. This section analyzes how the production of nanofluids has enabled the application of nanotechnology in the oil and gas industry.

9.1 GEOLOGICAL EXPLORATION AND PETROLEUM ACCUMULATION

Geological exploration is the initial step in the process of extracting petroleum, as it aims to locate formations that have the potential to contain oil. Geologists employ a range of tools and methods, such as seismic surveys, gravity and magnetic surveys, and geological mapping, to identify locations that have the capacity to accumulate oil. Source rocks, reservoir rocks, and trap structures are crucial geological elements that impact the buildup of petroleum.

Nanofluids, consisting of nanoparticles dispersed in a base fluid, have the ability to greatly improve geological exploration and the accumulation of petroleum. Nanofluids can be employed in these domains in the following manner:

- **Enhanced imaging techniques:** Nanoparticles distributed in drilling fluids can function as contrast agents in imaging techniques such as magnetic resonance imaging (MRI) and seismic surveys. Seismic survey fluids can be improved by adding nanoparticles, which can function as contrast agents to enhance the precision and correctness of seismic imaging. Geophysicists can enhance the visibility of subsurface features by infusing fluids containing nanoparticles into boreholes or reservoirs, which increases the contrast between different geological strata. This can result in improved characterization of geological formations, fault lines, and probable hydrocarbon sources.
  - Nanoparticles, when disseminated in seismic survey fluids, have the ability to decrease both friction and attenuation of seismic waves as they propagate through the subsurface. This enhances the transmission of signals and minimizes signal attenuation, leading to the production of seismic data that is more distinct and dependable. In addition, nanofluids can reduce the impact of scattering and absorption of seismic waves by limiting interactions between particles and fluids, resulting in improved signal-to-noise ratios and greater data interpretation. Nanofluids have the potential to enhance the signal-to-noise ratio of seismic data by diminishing background noise and interference. When nanoparticles are spread out in the fluid, they can function as acoustic absorbers or reflectors, reducing undesired noise and improving the signal from reflections beneath the surface. Geophysicists can use this technique to identify seismic occurrences with smaller amplitudes and obtain more precise and accurate information on structures beneath Earth's surface. Geoscientists can improve the quality of imaging investigations and better understand underground formations by injecting fluids containing nanoparticles into boreholes or reservoirs. This method allows for more precise identification of probable oil-rich areas.
  - Different types of nanofluids can be employed to enhance seismic surveys by OPTIMIZING fluid characteristics and signal transmission in the subsurface. Several often-employed nanofluids in this particular setting comprise:
    - **Silicon nanofluids**, namely silicon dioxide (SiO_2) nanoparticles, are extensively utilized in seismic surveys due to their exceptional acoustic characteristics and stability. Utilizing silica nanofluids can enhance the viscosity, density, and acoustic impedance of fluids, hence improving the transmission and reflection of signals in the subsurface. In addition, silica nanoparticles have the ability to decrease friction and the weakening of seismic waves, resulting in improved clarity and dependability of data collection.

- **Iron oxide nanofluids**, specifically magnetite (Fe_3O_4) or hematite (Fe_2O_3) nanoparticles, are commonly employed as contrast agents in seismic surveys because of their robust magnetic and acoustic characteristics. Iron oxide nanofluids have the ability to increase the distinction in signal between various geological layers and boost the precision of seismic imaging. In addition, iron oxide nanoparticles have the ability to function as acoustic absorbers, which results in the reduction of ambient noise and the enhancement of the signal-to-noise ratio of seismic data.
- **CNT nanofluids**, consisting of CNTs, have great potential for improving seismic surveys as a result of their distinctive mechanical and acoustic characteristics. Nanofluids containing CNTs have the potential to enhance the flow characteristics of fluids and decrease the resistance to motion in boreholes. This can result in improved fluid flow and enhanced transmission of signals. In addition, CNTs have the ability to improve the stability of fluids and prevent the settling of particles, therefore ensuring a consistent and even distribution in the subsurface.
- **Polymer nanofluids**, which consist of nanoparticles of substances like polyethylene glycol (PEG) or polyvinylpyrrolidone (PVP) dispersed in a polymer solution, are employed to enhance the viscosity and stability of fluids in seismic surveys. Polymer nanofluids have the ability to augment fluid density and acoustic impedance, hence improving the transmission and reflection of signals in the subsurface. In addition, polymer nanoparticles have the ability to create a viscoelastic network inside the fluid, which results in a decrease in both friction and the weakening of seismic waves.
- **Graphene nanofluids**, composed of single-layer carbon atoms organized in a hexagonal lattice, possess remarkable mechanical, electrical, and acoustic characteristics that make them well-suited for seismic investigations. Graphene nanofluids enhance fluid conductivity, stiffness, and acoustic impedance, resulting in improved signal transmission and reflection in the subsurface. Moreover, the incorporation of graphene nanoparticles can augment the stability of fluids and diminish friction, leading to a more streamlined fluid flow and enhanced data gathering.
- **Hybrid nanofluids**, which are composed of several nanoparticles dispersed in a base fluid, have synergistic benefits for improving seismic surveys. Hybrid nanofluids, which are created by combining nanoparticles with different qualities, such as silica for improving sound and iron oxide for enhancing MRI, offer a versatile approach to improving the performance of fluids and the quality of signals in the subsurface.
- The selection of nanofluid to improve seismic surveys is contingent upon various elements including the required characteristics of the fluid, geological circumstances, and survey goals. Researchers and industry experts can enhance data quality, resolution, and interpretation in geological exploration and resource evaluation by choosing the right nanofluid formulation.

- **Nanofluid tracers:** Nanoparticles serve as tracers for monitoring and managing reservoirs. Geoscientists can monitor the flow of fluids in reservoirs and evaluate how connected the reservoirs are by labeling nanoparticles with distinct chemical markers or isotopes. This data can assist in optimizing production techniques, maximizing the extraction of oil, and reducing the likelihood of reservoir compartmentalization. Nanofluid tracers have the potential to greatly improve geological exploration and petroleum accumulation processes. They can provide crucial information on the movement of fluids, connectivity within reservoirs, and dynamics of production. Nanofluid tracers can be employed in several fields in the following manner:
- Nanofluid tracers can be introduced into reservoirs to monitor and manage the flow of fluids during production and injection activities. Geoscientists can track the movement of fluids, determine how connected reservoirs are, and measure how well a reservoir is being swept by using nanoparticles that have been labeled with certain chemical markers or isotopes. This information is essential for optimizing production techniques, maximizing the extraction of oil, and limiting the potential for reservoir compartmentalization.
- Nanofluid tracers are useful for detecting locations within reservoirs where oil has been bypassed or not fully swept. Geoscientists can identify possible untapped oil deposits by injecting fluids containing tracers into injection wells and tracking their movement in production wells. This allows them to find locations where the fluid is not spreading efficiently or is being bypassed, suggesting regions with poor sweep efficiency. This enables the implementation of focused treatments, such as well reconfiguration or enhanced oil recovery techniques, to enhance oil output from these specific areas.
- Nanofluid tracers can be used to analyze and understand fluid flow patterns and variations in reservoir composition. Geoscientists can deduce flow velocities, preferential flow routes, and reservoir compartmentalization by examining the dispersion of tracer nanoparticles in production fluids over time. This data assists in the analysis of reservoir properties and facilitates the identification of geological elements such as faults, fractures, and obstacles that impact the flow and distribution of fluids.
- Nanofluid tracers are useful for evaluating the efficiency of individual wells and groups of wells in a reservoir. Operators can assess well-to-well communication, detect high-permeability channels or thief zones, and enhance well placement and spacing for optimal oil recovery by injecting fluids containing tracers into designated wells and monitoring the breakthrough of tracers in surrounding production wells. This enables more effective reservoir management and optimization of production.
- Nanofluid tracers are useful in identifying the movement of fluids across distinct reservoirs or geological formations. Geoscientists can track the movement of fluids across geological boundaries, assess the compartmentalization of reservoirs, and evaluate the potential for fluid communication or interference between adjacent reservoirs by injecting fluids containing

tracers into one reservoir or formation and monitoring the breakthrough of tracers in nearby wells or observation points.

- Nanofluid tracers are effective for monitoring and cleaning up polluted areas caused by operations related to the exploration and production of petroleum. Researchers can monitor the movement and dispersion of pollutants, evaluate the effectiveness of remediation efforts, and observe the long-term environmental effects by injecting fluids containing tracers into soil or groundwater that is contaminated with hydrocarbons or other pollutants. This facilitates more efficient remediation approaches and aids in reducing environmental hazards linked to oil and gas activities.

Various nanofluid tracers are used in geological exploration and petroleum accumulation to monitor the movement of fluids, connectivity of reservoirs, and dynamics of production. There are several prevalent categories of nanofluid tracers:

- **Metal-based nanoparticles**, including gold nanoparticles, silver nanoparticles, and iron oxide nanoparticles, are frequently employed as tracers in geological and petroleum engineering applications. These nanoparticles can be modified with distinct chemical markers or isotopes to facilitate the monitoring of fluid flow in reservoirs. Metallic nanoparticles provide exceptional stability, biocompatibility, and detection sensitivity, rendering them well-suited for use as tracers in intricate subsurface settings.
- **Polymer nanoparticles**, such as those made of polystyrene or PEG, serve as tracers to monitor fluid movement in porous materials. These nanoparticles can be manipulated to display particular surface characteristics and their ability to disperse, enabling accurate regulation of their actions in reservoirs. Polymer nanoparticles possess several benefits, including exceptional stability, adjustable surface chemistry, and resilience against chemical degradation. These qualities make them well-suited for conducting long-term monitoring and reservoir characterization studies.
- **Fluorescent nanoparticles**, such as quantum dots or semiconductor nanocrystals, are used as tracers in optical imaging to visualize fluid movement in reservoirs. These nanoparticles provide luminescent signals when stimulated by external light sources, allowing for the non-invasive and real-time observation of fluid flow routes and reservoir dynamics. Fluorescent nanoparticles possess several benefits, including heightened sensitivity, the ability to perform multiple tasks simultaneously, and compatibility with optical imaging methods. As a result, they are highly effective tools for monitoring and controlling reservoirs.
- **Radioactive nanoparticles**, such as gold nanoparticles labeled with radioisotopes or silica nanoparticles labeled with radioisotopes, are employed as tracers to monitor the flow of fluids and the connectivity of reservoirs by nuclear imaging techniques. The nanoparticles are labeled with radioactive isotopes that produce gamma rays or positrons, enabling the accurate detection and measurement of tracer concentrations in reservoir fluids. Radioactive nanoparticles possess several benefits, including exceptional

detection sensitivity, minimal background noise, and compatibility with quantitative imaging techniques. As a result, they are well-suited for accurately characterizing reservoirs and tracking fluids.

- **Magnetic nanoparticles**, specifically iron oxide or magnetite nanoparticles are utilized as tracers to monitor fluid flow in reservoirs through the use of MRI or magnetic resonance spectroscopy (MRS) techniques. These nanoparticles can be modified with magnetic tags or contrast ants to facilitate the accurate detection and visualization of tracer concentrations in reservoir fluids. Magnetic nanoparticles possess several benefits, including superior differentiation ability, the ability to perform imaging without causing harm, and compatibility with MRI/MRS equipment. As a result, they serve as significant resources for reservoir surveillance and monitoring.
- **Carbon-based nanoparticles**, such as CNTs or graphene oxide nanoparticles, are employed as tracers to monitor the movement and transportation of fluids in porous materials. These nanoparticles can be intentionally designed to display particular surface characteristics and their ability to disperse, enabling precise control over their actions in reservoirs. Carbon-based nanoparticles have several benefits, including exceptional mechanical durability, chemical resistance, and the ability to function well with various fluid systems. These qualities make them highly ideal for a broad spectrum of tracer uses in geological and petroleum engineering.
- In general, the selection of a nanofluid tracer is influenced by various aspects, including the desired qualities of the tracer, the level of sensitivity required for detection, the imaging method being used, and the conditions of the reservoir. Researchers and industry experts can improve the performance of tracers and gain a better knowledge of fluid behavior and reservoir dynamics in geological exploration and petroleum accumulation processes by carefully choosing the right nanofluid tracer formulation.

9.2 DRILLING OPERATIONS

After identifying a possible petroleum reservoir through geological surveys and exploration, drilling operations begin to reach the underground formations. Contemporary drilling methods have developed to adapt to different geological conditions and reservoir features. Multiple drilling techniques have been devised to accommodate diverse geological conditions encountered in the process of oil and gas exploration and production. The range of methods includes both basic techniques utilized in conventional reservoirs and more advanced ways necessary for demanding conditions, such as deepwater or unconventional formations. The following are several distinct drilling techniques:

Rotary drilling is the predominant technique employed for drilling oil and gas wells (**Figure 9.2**). The process entails the rotation of a drill bit, which is fastened to the lower end of a drill string, to form a hole in the subsurface of Earth. The rotary drive is positioned at the top of the wellbore, while the drilling bit is positioned at the bottom. The drill string serves as the mechanical connection between the two. Drilling bits are cutting tools utilized to create cylindrical holes. Depending on the specific drilling conditions, such as the type of formation being drilled, the

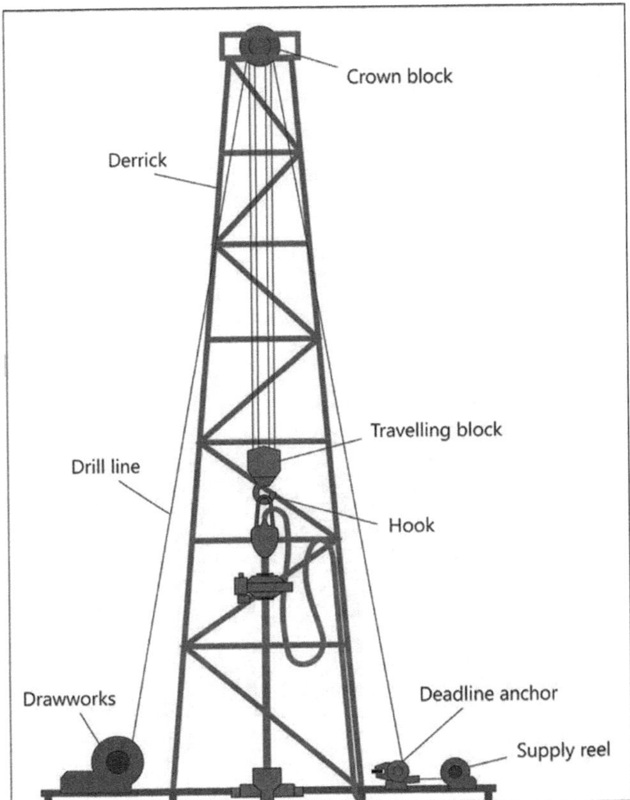

FIGURE 9.2 Schematic diagram representing the components of rotary drilling rig.[1]

bits are placed at the bottom of the drill string. As illustrated in **Figure 9.3** drill bit penetrates rock formations, while drilling mud is circulated to both cool the bit and transport rock cuttings to the surface. Rotary drilling is a highly adaptable method that can be modified to fit different geological conditions, making it well-suited for both onshore and offshore drilling operations.

Directional drilling is a method that enables operators to drill wells at various angles or horizontally to reach reservoirs that are located below impediments or at significant depths. These methods encompass deviated drilling, which involves purposely drilling the wellbore at an angle away from the vertical, and horizontal drilling, which entails drilling the wellbore parallel to the reservoir formation. Directional drilling allows for the extraction of resources from reservoirs situated beneath urban centers, environmentally delicate regions, or offshore platforms.

Extended reach drilling (ERD) refers to the process of drilling wells that are either horizontal or substantially deviated, having extended horizontal sections. The objective of ERD is to optimize reservoir exposure and increase production rates. ERD techniques enable operators to access remote or hard-to-reach reservoirs from a single drilling site, hence decreasing the requirement for multiple well pads and

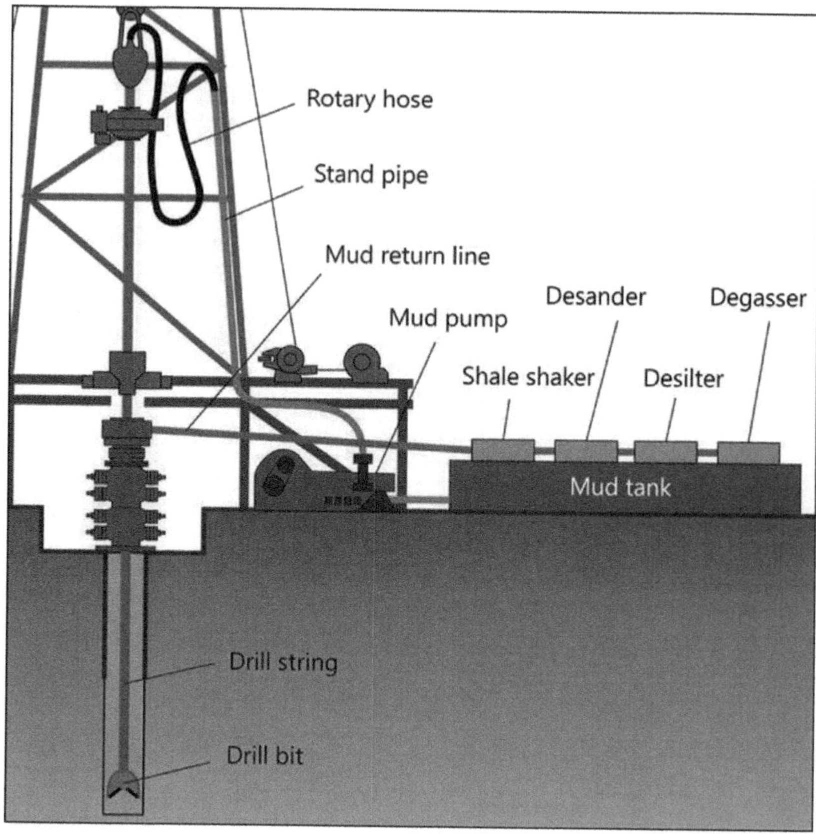

FIGURE 9.3 Schematic diagram representing the components of the rotary drilling operation.[1]

lessening the environmental footprint. ERD is a frequently employed technique in offshore drilling operations to reach and extract oil or gas from deepwater reserves. This method is utilized from either fixed or floating platforms.

The effectiveness of the **drilling fluid** is crucial for the success of a drilling operation. When drilling for oil and gas, a hole is dug from the surface to the reservoir, which can be located several kilometers away. To achieve this, a drilling bit is connected to a long string of drill pipe. The bit is weighted and rotated, breaking the rock into smaller pieces called cuttings. The drilling fluid is then circulated from the surface to the bit face through the drill pipe. The cuttings are lifted to the surface and separated from the drilling fluid using machinery. This machinery is responsible for lifting the cuttings. The fluid is then pumped back into the wellbore with the help of powerful pumps.

9.2.1 Composition of Drilling Fluids

- Base fluids, which can be water, nonaqueous, or pneumatic
- Solids, which can be active or inactive (inert)
- Additives, which are used to preserve the desired qualities of the system

9.2.2 Main Function of Drilling Fluids

- Remove any debris or sediment from the rock formation directly below the drill bit to collect rock cuttings.
- Extract the drilling tool from the well.
- Manage the pressures encountered during drilling operations and ensure the stability of the wellbore.
- Temporarily halt and then let go of plant cuttings.
- Apply a sealant to impermeable strata to minimize excessive loss of drilling mud.
- Reduce reservoir damage by employing reservoir drill-in fluid.
- Ensure that the bit and drilling assembly are adequately cooled, lubricated, and cleaned.
- Transfer hydraulic power to the downhole assembly.
- Ensure thorough assessment of formation characteristics.
- Manage and prevent corrosion.
- Enable measurement in the wellbore (measurement while drilling, logging while drilling).
- Enable the process of solidifying and finishing.
- Reduce the ecological footprint.

9.2.3 Types of Drilling Fluids

As shown in **Figure 9.4**, there are three main types of drilling fluids.

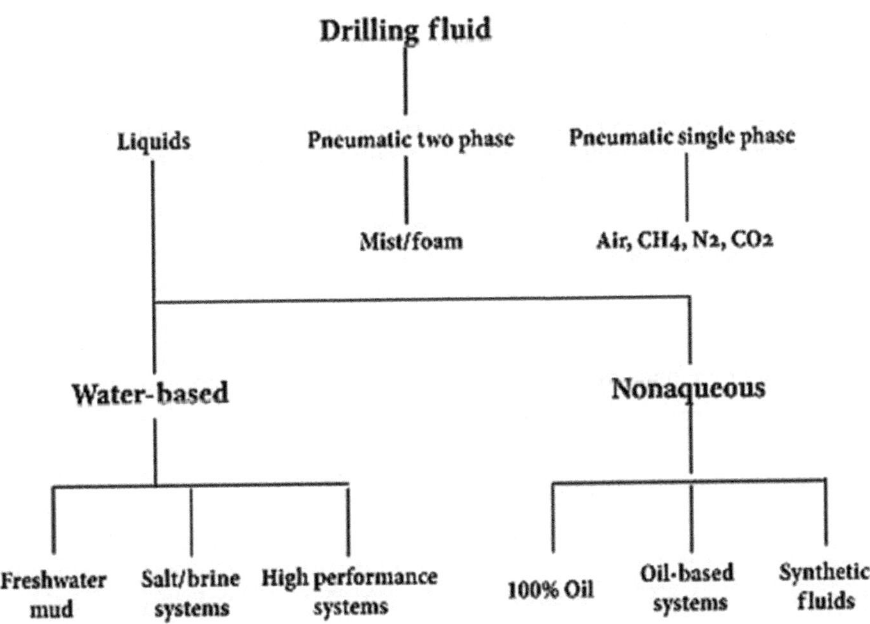

FIGURE 9.4 Types of drilling fluids.[2]

- **Water-based muds**, often known as WBM, are a type of drilling fluid that primarily consists of water. Particles of a solid nature are dispersed in water or brine. Oil can form an emulsion with water, in which case water is referred to as the continuous phase.
- **Oil-based drilling fluids** are drilling fluids that do not contain water. The oil contains suspended solid particles. The oil is emulsified with brine water or another low-activity liquid, where the oil acts as the continuous phase.
- **Pneumatic systems:** Drill cuttings are eliminated with the use of a forceful flow of air, natural gas, nitrogen, carbon dioxide, or another fluid that is introduced in a gaseous state. When there is a small amount of water entering the system, it is transformed into a gas or liquid form, such as mist or foam.

9.2.4 Limitations of Drilling Fluids

To explore new petroleum fields in challenging subsurface environments characterized by high-temperature and high-pressure (HT/HP) or low-temperature and low-pressure (LT/LP) conditions, it is necessary to develop and utilize innovative drilling fluids. These fluids must retain their rheological and filtration loss properties even in extremely hostile conditions.

9.2.5 Drawbacks

- Poor borehole stability
- Inadequate efficiency in the transportation of cut materials
- Inadequate lubrication capabilities
- Loss of drilling fluid

9.2.6 Solution

The drilling industry has the potential to significantly benefit from the implementation of nanotechnology. One fascinating possibility is to use nanoparticles (NPs) in drilling fluids to create intelligent drilling fluids that can perform optimally in a wide range of operating scenarios. Moreover, the potential to produce nanoparticles with customized specifications would have a substantial impact on the advancement of drilling fluids based on nanotechnology. This is because it will be feasible to produce customized fluids that can meet the specific needs of each operator to address various unique circumstances. Hence, the use of nanoparticles in the composition of advanced drilling fluids has the promise of resolving the present and future technological challenges encountered by the drilling sector.

9.3 NANOFLUID-BASED DRILLING FLUIDS

Various types of nanoparticles can be utilized in the production of nanofluids for drilling fluids. Each individual nanoparticle possesses unique traits and benefits. The following is a compilation of commonly employed nanoparticles in drilling fluid nanofluids:

Alumina (Al_2O_3), silica (SiO_2), and titanium dioxide (TiO_2) are often employed as metal oxide nanoparticles in drilling fluid nanofluids. Additional illustrations encompass titanium dioxide (TiO_2), among others. These nanoparticles have the capacity to enhance:

- Lubrication
- Thermal conductivity
- The integrity of wellbore connections

Carbon nanoparticles: Carbon nanoparticles, such as graphene, CNTs, and carbon black, have excellent lubricating properties and can help reduce friction between the drill string and the borehole wall. Graphene is a carbon nanoparticle. Furthermore, their exceptional thermal conductivity makes them very suitable for efficient heat management in drilling operations.

Clay nanoparticles, such as montmorillonite (MMT-$(Na,Ca)_{0.3}(Al,Mg)_2Si_4O_{10}(OH)_{2n}(H_2O)$) and bentonite, are commonly employed in water-based drilling fluids. The process of natural weathering in caves leads to the concentration and alteration of aluminosilicates that were originally present in the bedrock. Water causes MMT to undergo significant swelling and expansion compared to other clay forms. MMT's beneficial properties make it well-suited for various purposes, such as its use in the oil drilling industry as a constituent of drilling fluid during drilling operations. MMT, as a constituent of the smectite group, is a type of clay with a 2:1 structure. It is composed of two layers of silica tetrahedra sandwiched between two layers of alumina octahedra at the core of the clay as shown in **Figure 9.5**.

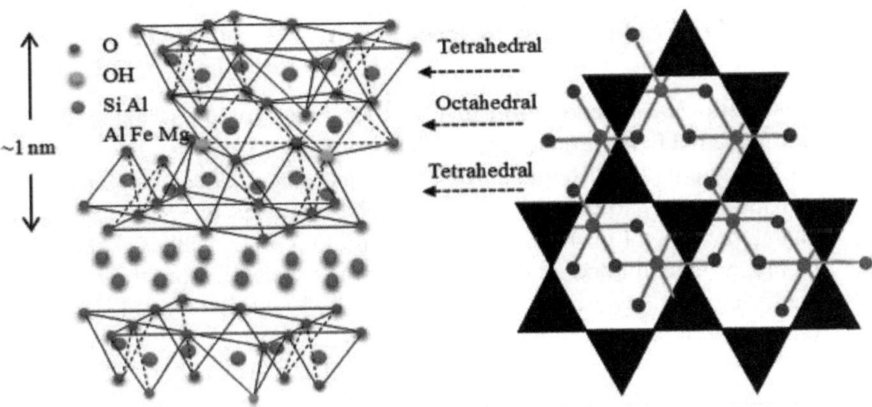

FIGURE 9.5 Structure of montmorillonite (MMT) clay.[3]

By employing these nanoparticles, it is possible to enhance rheological properties, enhance wellbore stability by creating a filter cake, and achieve more efficient control over fluid loss.

Metal nanoparticles, including copper (Cu), silver (Ag), and gold (Au), can be included in nanofluids used in drilling fluid to provide antibacterial properties. This results in a decrease in the proliferation of microorganisms and the formation of biofouling in the wellbore.

Polymer nanoparticles, such as PEG and polyacrylamide, can enhance fluid rheology, decrease fluid loss, and improve wellbore stability. They are particularly suitable for use in oil-based drilling fluids.

Functionalized nanoparticles are nanoparticles that have been altered by adding certain chemical groups or coatings to provide desired characteristics, such as compatibility with the base fluid, stability, and targeted functionality (such as controlling fluid loss or inhibiting corrosion). These nanoparticles can be intentionally engineered to demonstrate these specific characteristics.

Hybrid nanoparticles, consisting of various combinations of materials such as metal oxide-carbon hybrids, have the capacity to produce synergistic effects. These nanoparticles amalgamate the attributes of individual nanoparticles to enhance their efficacy in drilling fluids.[4]

Various nanoparticles can be utilized in the composition of nanofluids for drilling fluids, and a few examples are provided here. These examples are merely a small selection in total. When choosing nanoparticles, factors such as cost-effectiveness and environmental impact are also considered. These features encompass the distinct challenges encountered during drilling operations, the essential characteristics needed in the drilling fluid, and the necessary parameters to be met.[5]

9.4 FEW MORE APPLICATIONS OF NANOFLUIDS IN THE FIELD OF PETROLEUM SCIENCE AND TECHNOLOGY

- **Completion and production** involve the process of preparing a drilled well to be ready for production when it has reached the desired depth. This process entails the installation of casing and its subsequent cementing to provide stability to the wellbore. Additionally, the casing is perforated to enable the flow of oil from the reservoir rock into the wellbore. After finishing, production operations begin to extract crude oil from the reservoir and bring it to the surface. Primary production is dependent on the natural pressure in reservoirs to force oil to the surface. Typically, pump jacks or sucker rod pumps are employed to elevate oil from the wellbore. Secondary recovery methods, such as water flooding or gas injection, can be used to improve oil recovery by shifting the residual oil toward producing wells. Enhanced oil recovery methods, such as thermal, chemical, and mechanical approaches, are employed to augment oil production from mature or difficult sources.[6]

- **Surface facilities and transportation** involve the transfer of extracted crude oil to processing and treatment facilities. This involves the process of segregating the oil from water and gas, eliminating any impurities and pollutants, and ensuring the oil is in a stable condition for transportation. After undergoing processing, the unrefined oil is sent through pipelines, tankers, or alternative methods to refineries or other locations for additional processing and purification into different petroleum products. The process of extracting petroleum is an intricate and interdisciplinary undertaking that encompasses geological exploration, drilling engineering, and reservoir management. Comprehending the geological elements that impact the accumulation of petroleum, utilizing sophisticated drilling techniques, and executing effective production methods are crucial for achieving success in oil extraction operations. To meet the increasing global energy demand while reducing environmental harm, it is essential to focus on the advancement of inventive technology and sustainable methods for petroleum production.[7]

9.5 CHALLENGES IN UTILIZING NANOFLUIDS IN THE FIELD OF PETROLEUM SCIENCE AND TECHNOLOGY

Although nanofluids possess the capability to enhance various aspects of petroleum research and technology, their extensive adoption is impeded by several barriers, which include:

- Attaining and sustaining a stable dispersion of nanoparticles in base fluids such as water or oil might pose challenges. This is an obstacle that can be surmounted. Over time, nanoparticles tend to aggregate due to van der Waals forces or electrostatic interactions. As a consequence, sedimentation occurs and their efficiency diminishes. To overcome this challenge, it is imperative to create effective dispersants and techniques for stabilization.[8]
- **Compatibility and interaction**: Nanoparticles might potentially interact with other components in the drilling fluid or reservoir fluids, leading to unwanted effects such as changes in rheological properties, fluid loss characteristics, or formation damage. To ensure compatibility and prevent adverse consequences, it is crucial to possess a comprehensive understanding of the interactions that take place between nanoparticles and other chemicals.[9]
- Incorporating nanoparticles into drilling fluids can significantly increase costs compared to the use of conventional fluids. The high cost of synthesizing, purifying, and functionalizing nanoparticles limits their potential applications. Developing cost-efficient production processes and optimizing nanoparticle compositions are crucial for enhancing cost-effectiveness.
- **Issues related to scaling up**: The transition from laboratory-scale operations to field-scale activities in the manufacturing of nanofluids poses

several technological and logistical challenges. To achieve consistent dispersion of nanoparticles, ensure quality control, and successfully incorporate nanofluid handling systems into current drilling processes, meticulous design and substantial investment in infrastructure are necessary.

- Concerns pertaining to health, safety, and the environment nanoparticles can pose possible risks to the health and safety of workers when they are handled or exposed to. Furthermore, there is a legitimate concern regarding their environmental impact, encompassing potential toxicity and the enduring repercussions on ecosystems. To ensure responsible usage of nanofluids and address these problems, it is crucial to establish and enforce appropriate safety protocols, conduct thorough risk assessments, and employ effective environmental monitoring systems.[10]
- Regulatory permission and compliance requirements may be necessary for the utilization of nanofluids in petroleum operations. This is due to the possibility of noncompliance with regulations. Regulatory agencies may require a thorough assessment of the toxicity, environmental impact, and effectiveness of nanoparticles in drilling fluids or reservoir treatments before issuing approval for their use. Ensuring adherence to regulatory mandates and obtaining the necessary authorizations can be a significant burden in terms of time and resources.
- **An incomplete comprehension of nanoparticle behavior**: Although extensive studies have been carried out, there is still a lack of understanding regarding certain aspects of nanoparticle behavior in complicated petroleum environments. To enhance the efficiency of nanofluid formulations and accurately forecast their performance, further investigation is necessary to explore issues such as the transportation, retention, and reactivity of nanoparticles in porous materials.[11]
- Interdisciplinary collaboration among scientists, engineers, regulators, and industry stakeholders is necessary to progress nanofluid technology and fully realize its potential benefits in petroleum research and technology. This is necessary to overcome the challenges that have previously arisen. To successfully utilize nanofluids in petroleum applications, it is crucial to sustain research and innovation efforts and promote information sharing. This will ensure the safe and efficient usage of nanofluids in this context.[12]

9.6 SUMMARY

The use of nanofluids in petroleum research and technology presents excellent prospects for improving drilling, production, and reservoir management procedures. Nanofluids are colloidal suspensions that consist of nanoparticles dispersed in base fluids. These nanofluids have the ability to enhance the properties of petroleum fluids and formations in several advantageous ways. These benefits encompass enhanced heat transfer, less resistance, improved control over fluid flow characteristics, increased stability of the wellbore, and the capacity to address drilling difficulties such as lost circulation.

Although nanofluids have the potential to bring numerous advantages, there are various obstacles that hinder their broad use in petroleum applications. These tasks involve ensuring the stability of dispersion; resolving compatibility and interaction problems; maximizing cost efficiency; overcoming challenges in scaling up; addressing health, safety, and environmental issues; complying with regulatory requirements; and enhancing knowledge of nanoparticle behavior in complicated petroleum environments.

Despite the presence of obstacles, the deliberate use of nanofluids in petroleum exploration, production, and reservoir management has great potential for raising efficiency, cutting expenses, and promoting sustainability. Nanofluids have the potential to significantly impact the petroleum sector in the future due to continuous developments and collaborative efforts.

REFERENCES

1. Paulauskiene, T. (2018). Petroleum extraction engineering. In *Recent Insights in Petroleum Science and Engineering*. InTech. http://doi.org/10.5772/intechopen.70360
2. Caenn, R., Darley, H. C. H., & Gray, G. R. (2017). Chapter 1—Introduction to drilling fluids. In Caenn, R., Darley, H. C. H., & Gray, G. R. (Eds.), *Composition and Properties of Drilling and Completion Fluids* (7th ed., pp. 1–34). Gulf Professional Publishing. ISBN 9780128047514. https://doi.org/10.1016/B978-0-12-804751-4.00001-8.
3. Yaghmaeiyan, N., Mirzaei, M., & Delghavi, R. (2022). Montmorillonite clay: Introduction and evaluation of its applications in different organic syntheses as catalyst: A review. *Results in Chemistry*, *4*, 100549. https://doi.org/10.1016/j.rechem.2022.100549
4. Khedkar, R. S., Sonawane, S. S., & Wasewar, K. L. (2014). Heat transfer study on concentric tube heat exchanger using TiO2–water based nanofluid. *International Communications in Heat and Mass Transfer*, *57*, 163–169.
5. Vijay, J., & Sonawane Shriram, S. (2015). Investigations on rheological behaviour of paraffin based Fe_3O_4 nanofluids and its modelling. *Research Journal of Chemistry and Environment*, *19*(12), 22–29.
6. Sonawane, S. S., Khedkar, R. S., & Wasewar, K. L. (2015). Effect of sonication time on enhancement of effective thermal conductivity of nano TiO2–water, ethylene glycol, and paraffin oil nanofluids and models comparisons. *Journal of Experimental Nanoscience*, *10*(4), 310–322.
7. Khedkar, R. S., Shrivastava, N., Sonawane, S. S., & Wasewar, K. L. (2016). Experimental investigations and theoretical determination of thermal conductivity and viscosity of TiO2–ethylene glycol nanofluid. *International Communications in Heat and Mass Transfer*, *73*, 54–61.
8. Nishant, K., & Sonawane, S. S. (2016). Influence of CuO and TiO2 nanoparticles in enhancing the overall heat transfer coefficient and thermal conductivity of water and ethylene glycol based nanofluids. *Research Journal of Chemistry and Environment*, *20*, 8.
9. Sonawane, S. S., & Juwar, V. (2016). Optimization of conditions for an enhancement of thermal conductivity and minimization of viscosity of ethylene glycol based Fe_3O_4 nanofluid. *Applied Thermal Engineering*, *109*, 121–129.
10. Kumar, N., & Sonawane, S. S. (2016). Experimental study of Fe2O3/water and Fe2O3/ethylene glycol nanofluid heat transfer enhancement in a shell and tube heat exchanger. *International Communications in Heat and Mass Transfer*, *78*, 277–284.
11. Kumar, N., Sonawane, S. S., & Sonawane, S. H. (2018). Experimental study of thermal conductivity, heat transfer and friction factor of Al_2O_3 based nanofluid. *International Communications in Heat and Mass Transfer*, *90*, 1–10.
12. Kumar, N., Urkude, N., Sonawane, S. S., & Sonawane, S. H. (2018). Experimental study on pool boiling and Critical Heat Flux enhancement of metal oxides based nanofluid. *International Communications in Heat and Mass Transfer*, *96*, 37–42.

10 Progress and Challenges for Hybrid Nanofluids' Future Prospects

10.1 OPPORTUNITIES AND CHALLENGES IN THE APPLICATION OF NANOFLUIDS AND HYBRID NANOFLUIDS

Hybrid nanofluids are a significant improvement in the field of fluid engineering. In comparison to regular nanofluids, hybrid nanofluids have the potential to exhibit improved thermal, rheological, and other properties. The intricacy of these systems, on the contrary, brings forth a variety of difficulties that need to be carefully addressed. In this exposition, we will investigate the reasons why hybrid nanofluids present more difficulties than nanofluids. We will investigate synthesis, optimization, stability, compatibility, characterization, and the interdisciplinary nature of resolving these complexities,

- **Complexity of the synthesis:**
  To produce hybrid nanofluids, it is necessary to incorporate a variety of nanoparticles or additives into base fluids. Because hybrid nanofluids involve careful consideration of the interactions between distinct nanoparticles and their compatibility with the base fluid, in contrast to single-component nanofluids, which are primarily concerned with the uniform dispersion of a single type of nanoparticle, hybrid nanofluids are not as straightforward. It is necessary to modify synthesis methods to accommodate many components, which results in an increase in the complexity of the process. When it comes to dispersing a single type of nanoparticle, for instance, sonication or stirring might be efficient. However, to achieve uniform dispersion of numerous types of nanoparticles, these methods might require more advanced approaches.
- **Optimization challenges:**
  Enhanced properties can be achieved by the utilization of hybrid nanofluids, which have the ability to provide synergistic effects between various nanoparticles. Nevertheless, it can be difficult to attain the necessary qualities by optimizing the combination and concentration of nanoparticles because of the complexity involved. To obtain the appropriate level of performance, it is necessary to carefully balance the interactions that occur between the various types of nanoparticles and the interactions that these nanoparticles have with the base fluid. It is necessary to conduct exhaustive experiments and optimizations to do this, taking into consideration aspects such as particle size, surface chemistry, and concentration.

DOI: 10.1201/9781003595137-10

- **Regarding long-term stability issues:**
  Maintaining the stability of hybrid nanofluids presents significant challenges. Increasing the number of different kinds of nanoparticles in a mixture makes it more likely that the particles may clump together or silt, which ultimately results in instability. The stability of the system can also be affected by the interactions that occur between the various nanoparticles and the base fluid. To effectively stabilize hybrid nanofluids, it is necessary to have a more in-depth understanding of the mechanisms that drive stability and to design tactics that can limit instability. It is possible that this will require the utilization of surfactants, surface modification techniques, or other stabilization approaches that are specifically tuned to the features of hybrid nanofluids.
- **Compatibility challenges between the nanoparticles:**
  There is a possibility that hybrid nanofluids will experience compatibility issues as a result of the interaction between various kinds of nanoparticles and the base fluid. It is possible for certain nanoparticles to display compatibility concerns with the base fluid or with each other, which can result in phase separation, aggregation, or chemical reactions that damage the properties of the fluid. To achieve compatibility between numerous components in a hybrid nanofluid, it is necessary to pick nanoparticles and additives with great care and to take into consideration how these components interact with the base fluid as well as with each other.
- **The complexity of the characterization:**
  When compared to the process of characterization of single-component nanofluids, the processes of characterization of hybrid nanofluids are more complicated. When attempting to characterize qualities such as thermal conductivity, viscosity, and stability, the existence of different types of nanoparticles adds an extra layer of complexity. It is possible that standardized characterization methodologies will need to be modified or invented expressly for hybrid nanofluids to appropriately evaluate the properties and performance of these technologies. It is also possible that the interpretation of characterization data will be more difficult due to the presence of several components and the possibility of interactions between those components.
- **The nature of interdisciplinarity:**
  To effectively address the issues posed by hybrid nanofluids, a multidisciplinary approach is required. Nanotechnology, fluid dynamics, materials science, chemistry, and engineering are some of the domains that researchers need to draw upon to develop appropriate solutions. The development of novel synthesis procedures, the proper characterization of hybrid nanofluids, and the comprehension of the underlying principles that govern their behavior are all impossible without the participation of collaborative efforts. The research process is made more difficult by the fact that it is interdisciplinary, but it is essential for overcoming the obstacles and achieving the promise of hybrid nanofluids.

Hybrid nanofluids are a viable path for improving fluid properties; yet, the complexity of these nanofluids presents major obstacles. Synthesis, optimization, stability, compatibility, characterization, and interdisciplinary collaboration are some of the major areas that need to be addressed to solve these obstacles. Researchers are able to unlock the full potential of hybrid nanofluids for a wide range of applications by addressing these complications. These applications include heat transfer and energy storage, as well as biomedical and environmental engineering.

10.1.1 Opportunities

Nanofluids and hybrid nanofluids are advanced developments in the field of fluid dynamics that show great potential for use in a wide range of sectors. These sophisticated fluids, created by scattering nanoparticles in base fluids, present various possibilities and also present certain difficulties that require investigation.

- Nanofluids demonstrate a substantial enhancement in heat conductivity when compared to their basic fluids. Nanoparticles, such as carbon nanotubes or metal oxides, improve heat transfer capacities, making them useful for applications such as cooling systems in electronics, automobile engines, and thermal management in industrial operations.
- Energy efficiency can be enhanced by using nanofluids in heat transfer systems because of their exceptional thermal characteristics. Nanofluids have the potential to decrease energy usage in heating and cooling applications by enhancing heat transfer rates. This can contribute to sustainability initiatives and result in cost savings.
- **Miniaturization**: The advancement of microfluidic systems and nanotechnology enables the manipulation of fluids at the nanoscale, which creates opportunities for the development of innovative tiny devices and systems. Nanofluids provide accurate manipulation of fluid characteristics, facilitating the creation of small and effective devices for healthcare, environmental monitoring, and microelectronics purposes.
- Nanofluids exhibit potential in biomedical applications such as medication administration, imaging, and hyperthermia treatment. Due to their exceptional characteristics, such as elevated thermal conductivity and surface area, they are well-suited for precise therapeutic and diagnostic applications, which have the potential to transform medical treatments and diagnostics.
- Nanoparticles can be added to lubricants to improve their lubricating qualities, resulting in reduced friction and wear in machinery and engines. As a result, the equipment has a longer lifespan, lower maintenance costs, and better performance, especially in demanding industries like aerospace and automotive.

10.1.2 Challenges

- Stability and agglomeration pose significant obstacles in the realm of nanofluids, since they hinder the dispersion of particles and promote their clumping together. Nanoparticles possess a significant amount of surface energy, which causes them to clump together gradually. This aggregation can have a negative impact on the characteristics and performance of fluids. It is essential to develop efficient stabilizing procedures to guarantee the long-term stability and constant performance of nanofluids.

- **Expense**: The process of manufacturing nanoparticles and integrating them into base fluids can be financially burdensome, which hinders the extensive use of nanofluids, particularly in large-scale industrial settings. To make nanofluids economically feasible for different industries, it is necessary to employ cost-effective manufacturing processes and scalable production procedures.
- **Compatibility and material selection**: It is important to consider the compatibility between nanoparticles and base fluids, as well as the potential degradation of materials, to guarantee the dependability and safety of nanofluid systems. It is crucial to choose nanoparticles and base fluids that have suitable qualities and low reactivity to prevent any negative impact on equipment and processes.
- The absence of standardized procedures for measuring nanofluid characteristics and performance poses difficulties in comparing results across research and ensuring reproducibility. Developing standardized testing methods and characterization techniques is crucial for progressing research, aiding technology advancement, and encouraging industry adoption of nanofluids.
- Nanoparticles provide significant health and environmental hazards as a result of their diminutive size and distinctive characteristics. It is crucial to conduct a precise evaluation of their toxicity and environmental consequences to guarantee the implementation of safe handling and disposal methods. Enforcing appropriate safety protocols and regulations is essential for reducing the potential dangers linked to the utilization of nanofluids.
- Nanofluids and hybrid nanofluids have significant potential to transform multiple sectors by providing improved heat transfer, energy efficiency, and innovative capabilities. Nevertheless, it is crucial to tackle obstacles concerning stability, affordability, compatibility, standardization, and safety to fully reap their advantages and promote their broad use in practical contexts. Ongoing research and innovation are essential for surmounting these limitations and fully harnessing the potential of nanofluids in the future.

10.2 INDUSTRIAL OPERATION AND SCALE-UP CHALLENGES FOR NANOFLUID APPLICATIONS

The utilization of nanofluids and hybrid nanofluids in industrial processes shows great potential for a wide range of applications. However, there are certain obstacles that need to be overcome to ensure their successful implementation. Nanofluids are colloidal suspensions of nanoparticles in a base fluid that have been engineered. Hybrid nanofluids, on the contrary, are a combination of different types of nanoparticles or have additives added to them. Both nanofluids and hybrid nanofluids provide improved thermal conductivity, enhanced heat transfer rates, and other desirable properties when compared to traditional heat transfer fluids. Nevertheless, the implementation of these technologies in industrial settings has challenges pertaining to manufacturing, stability, cost, safety, and environmental considerations. This thorough investigation examines the obstacles and possible remedies in great detail.

10.2.1 Challenges with Nanofluid Production

i. **Dispersing nanoparticles:**
One of the primary challenges in industrial processes that use nanofluids is the attainment of homogeneous dispersion of nanoparticles within the base fluid. Agglomeration, settling, and inadequate dispersion may arise as a result of van der Waals forces, electrostatic interactions, and constraints imposed by Brownian motion. As a result, there are variations in the characteristics and diminished efficiency. It is crucial to develop strong dispersion techniques, such as ultrasonication, mechanical stirring, or surface modification of nanoparticles, to ensure stability and optimize performance.

ii. **Scalability:**
Scaling up the manufacturing of nanofluids from the laboratory to an industrial level presents considerable difficulties. Traditional methods of synthesis, such as chemical precipitation or sol–gel processes, may not be easily expandable or economically feasible. To achieve larger production volumes while ensuring consistent nanoparticle size distribution, purity, and stability, it is necessary to employ creative engineering techniques and optimize various process parameters.

10.2.2 Problems Related to Stability

i. **Sedimentation and agglomeration:**
The presence of nanoparticles in nanofluids has a tendency to gradually settle down, resulting in sedimentation and stratification, particularly in situations where there is no movement or very low fluid flow. The process of agglomeration worsens this problem by decreasing the actual surface area and the improvement in heat conductivity. To address stability difficulties, it is necessary to optimize the particle size, shape, and surface chemistry to reduce aggregation. Additionally, stabilizing chemicals or surfactants should be used.

ii. **Thermal decomposition:**
High temperatures and extended thermal cycling can lead to the deterioration of nanofluid stability. The surfactants or organic additions may undergo heat breakdown, resulting in phase separation or alterations in viscosity and thermal conductivity. It is crucial for industrial applications to develop formulations that are thermally stable and to investigate new stabilizers that can withstand heat degradation.

10.2.3 Cost Analysis

i. **Cost of nanoparticles:**
The exorbitant expense of nanoparticles, especially those possessing distinctive characteristics or customized capabilities, poses a substantial obstacle to their extensive use in industrial operations. Mass production methods and the benefits of producing on a large scale can lower expenses, but it is essential to develop inventive approaches for producing or reusing nanoparticles in a cost-efficient manner to improve affordability and competitiveness.

ii. **Costs of processing:**
In addition to the expenses related to nanoparticles, the supplementary costs linked to the synthesis, dispersion, and quality control of nanofluids also influence the total cost-effectiveness. Efficiently optimizing manufacturing processes, reducing energy usage, and simplifying production workflows are crucial for reducing processing expenses and enhancing the economic viability of large-scale industrial operations.

10.2.4 Safety and Environmental Considerations

i. **Hazards to health and safety:**
Workers may face health and safety hazards when handling, synthesizing, and disposing of nanofluids containing nanoparticles. Exposure to nanoparticles by inhalation or contact with the skin may result in respiratory issues, skin irritation, or enduring health consequences. It is crucial to implement strict safety regulations, engineering controls, and protective measures, as well as conduct extensive risk assessments, to guarantee occupational health and reduce potential dangers.

ii. **Environmental impact:**
An additional area of concern is the environmental ramifications associated with the manufacture, utilization, and disposal of nanofluids. The release or discharge of nanofluids into ecosystems can result in the introduction of nanoparticles into water bodies or soil, which might have ecological consequences and the ability to accumulate in the food chain. It is crucial to engage in the development of environmentally friendly nanofluid formulations, the investigation of biodegradable nanoparticles, and the implementation of efficient waste management systems to decrease environmental impacts and advance sustainability.

10.2.5 Enhancing Performance

i. **Enhancing heat transfer:**
Although nanofluids exhibit higher thermal conductivity than traditional heat transfer fluids, the primary goal in industrial operations is still to maximize heat transfer enhancement. By optimizing the concentration, size distribution, and dispersion quality of nanoparticles, as well as making

improvements to the system design, implementing flow enhancement techniques, and optimizing the heat exchanger, we may improve heat transfer performance and increase energy efficiency.

ii. **Operational stability:**

Nanofluid stability can be compromised by the harsh working conditions commonly seen in industrial processes, such as high temperatures, pressures, and mechanical stresses. To ensure strong stability in various operating conditions, it is necessary to conduct thorough performance testing, verify material compatibility, and carry out dependability studies. It is crucial to create nanofluid formulations that can withstand harsh circumstances to provide dependable and durable industrial use.

10.3 LIFE CYCLE ANALYSIS OF NANOFLUIDS

Life cycle analysis (LCA) is a methodical strategy used to evaluate the environmental effects linked to a product, process, or service across its complete life cycle, starting with the extraction of raw materials to the final disposal or recycling stage. Performing an LCA is crucial for comprehending the environmental consequences of different activities and guiding decision-making processes to advance sustainability and optimize resource utilization. This text provides a comprehensive examination of the significance of undertaking LCA, along with a detailed and systematic approach to carrying out an LCA.

10.3.1 Significance of Conducting LCA

A comprehensive assessment: LCA offers a comprehensive perspective on the environmental consequences linked to a product or process, taking into account all phases of its life cycle. This comprehensive assessment enables stakeholders to identify and prioritize areas with significant environmental impact, examines the trade-offs involved, and formulates effective policies for environmental management and enhancement.

LCA results are essential tools for businesses, governments, and consumers to support decision-making. They help guide product design, process optimization, investment decisions, and policy creation. LCA allows for the measurement and comparison of environmental effects and provides the necessary information for making well-informed decisions that take into account both environmental concerns and economic and social aspects.

Resource efficiency is a key benefit of LCA since it enables the identification of possibilities to reduce waste, minimize contamination, and optimize resource usage over the whole life cycle of products and activities. LCA helps sustainable resource management and circular economy concepts by optimizing material and energy flows, lowering resource consumption, and reducing environmental impacts.

LCA enables the recognition and evaluation of potential environmental hazards and consequences linked to various phases of a product's life cycle. Through the assessment of variables like as emissions, waste production, and toxicity, LCA aids in the reduction of risks, guaranteeing adherence to regulations, and minimizing negative environmental impacts.

LCA enhances openness and accountability through the use of a methodical and uniform framework for evaluating environmental impact. Through the documentation of data sources, assumptions, and methodology, the outcomes of LCA can be thoroughly examined, validated, and effectively conveyed to stakeholders. This practice fosters confidence, credibility, and accountability in decision-making processes.

Continuous Improvement: LCA functions as a mechanism for ongoing enhancement and originality, propelling sustainability endeavors, product advancement, and process streamlining. LCA promotes sustainability and encourages firms to continuously learn and innovate by identifying areas for environmental improvement and establishing reduction targets.

10.3.2 Procedure for Conducting LCA

10.3.2.1 Definition of Objectives and Scope

Specify the goals and limitations of the LCA study, which involve identifying the specific product or system to be evaluated, determining the functional unit of analysis, and selecting the environmental impact categories that are of interest. Define the limits of the system, which should encompass all stages of the life cycle such as raw material extraction, production, usage, and disposal. Also, identify any activities or impacts that are not included in the analysis.

10.3.2.2 Life Cycle Inventory (LCI)

Life cycle inventory (LCI) refers to the systematic collection and quantification of inputs, outputs, and environmental impacts associated with a product or process throughout its entire life cycle. Compile a thorough inventory of all inputs (such as materials, energy, and water) and outputs (such as emissions and waste) that are linked to each stage of the life cycle. Gather data on the utilization of resources, release of emissions, and production of trash from original sources, comprehensive literature reviews, databases, and consultations with experts.

10.3.2.3 Life Cycle Impact Assessment

Life cycle impact assessment is a method used to evaluate the potential environmental impacts of a product or process throughout its entire life cycle. Assess the possible environmental effects linked to the LCI data by utilizing impact assessment methodologies and models. Utilize characterization parameters to measure the magnitude of effects in many environmental impact categories, including climate change, resource depletion, acidification, and human toxicity.

10.3.2.4 Explanation

Analyze the results of LCA to identify regions with high environmental impact, determine the main factors contributing to these impacts, and pinpoint opportunities for improvement. Analyze the compromises between several environmental impact categories and take into account the uncertainty and sensitivity of LCA results.

10.3.2.5 Analysis of Improvement

Identify potential areas for environmental improvement and enhancement of sustainability based on the findings of an LCA. Create and implement plans and procedures to minimize the negative effects on the environment, maximize the efficient use of resources, and decrease potential risks at every stage of the product's life cycle.

10.3.2.6 Documentation and Dissemination of Information

Produce a report that clearly and thoroughly outlines the LCA approach, assumptions, data sources, and results in a manner that is easily understood and leaves no room for ambiguity. Effectively convey the findings of an LCA to stakeholders through reports, presentations, and other communication channels. Ensure that the communication is clear, accurate, and relevant to the decision-making processes.

10.3.2.7 Evaluation and Verification

Conduct a thorough examination and verification of the LCA research by subjecting it to peer review, gathering input from stakeholders, and implementing quality assurance procedures. To improve the credibility and reliability of LCA results, it is important to address any uncertainties, restrictions, or discrepancies that are detected throughout the review process. By adhering to these procedures, those with a vested interest can carry out a comprehensive and enlightening examination of the entire lifespan of a product or process. This study yields useful knowledge regarding the ecological consequences and aids in making informed choices, ultimately fostering sustainable progress.

NOTE

The fundamental principles and methodologies of LCA are applicable to both nanofluids and hybrid nanofluids. However, the evaluation of hybrid systems requires additional considerations regarding the integration of multiple types of nanoparticles, interactions between nanoparticles, and potential synergistic effects on environmental performance.

10.3.3 Case Study

CONDUCT CRADLE-TO-GRAVE LCA AND COMPARE LCA OF CuO/WATER NANOFLUID WITH CuO–ZnO/WATER HYBRID NANOFLUID

The comparison of LCAs between CuO/water nanofluid and CuO–ZnO/water hybrid nanofluid entails evaluating the environmental effects linked to each nanofluid formulation over their complete life cycles (**Figure 10.1**). The following is a comparative analysis focusing on major elements of their life cycle stages:

- **Extraction of raw materials and production of nanoparticles:**
  CuO nanoparticles: The procurement of CuO nanoparticles' raw materials, usually copper ore, entails laborious mining activities and chemical refinement that consume significant amounts of energy. This process is responsible for resource depletion, habitat damage, and the release of contaminants.

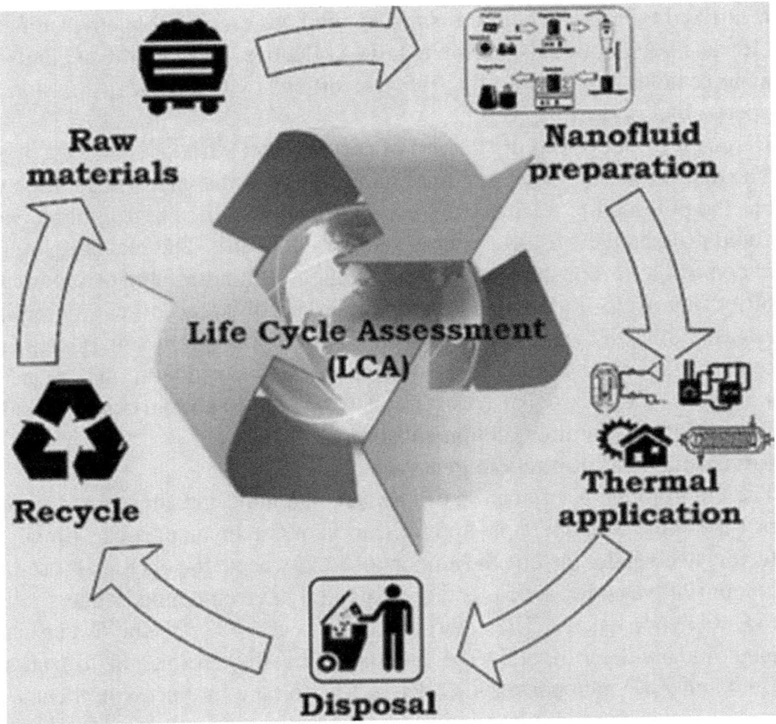

FIGURE 10.1 Illustration depicting the schematic design of life cycle assessment (LCA) applied to nanofluids.[1]

ZnO nanoparticles: The acquisition of raw materials for ZnO nanoparticles, including zinc ore, entails mining and processing operations that have environmental consequences. These consequences include energy usage, land disruption, and emissions.

- **Production of nanofluids**:
  The production of CuO nanofluid includes the dispersion of CuO nanoparticles into a water-based solution, which necessitates the use of energy-intensive dispersion processes and perhaps the addition of chemical additives for stabilization. This process is responsible for energy consumption, chemical utilization, and wastewater production.[2]

  The production of CuO–ZnO hybrid nanofluid entails the amalgamation of CuO and ZnO nanoparticles in a water-based solution, similar to the production process of CuO nanofluid. Furthermore, it is important to take into account the optimization of nanoparticle ratios, dispersion techniques, and the compatibility between CuO and ZnO nanoparticles.
- **Utilization of the product:**
  The CuO nanofluid is utilized in heat transfer systems to augment thermal conductivity and optimize energy efficiency. The environmental effects during usage are mostly linked to energy consumption, system functioning, and maintenance. The CuO–ZnO hybrid nanofluid, like the CuO nanofluid,

is utilized to improve heat transmission and has comparable environmental effects when applied. Nevertheless, the collective impacts of CuO and ZnO nanoparticles can potentially affect the efficiency of heat transmission and operational performance.[3]

- **Disposal or recycling at the end of the product's life:**
  The disposal of CuO/water nanofluid products at the end of their life cycle has the potential to release CuO nanoparticles into the environment, which could pose dangers to ecosystems and human health. The recycling of CuO nanoparticles is constrained by technological difficulties and economic viability. The CuO–ZnO hybrid nanofluid, when disposed of, can potentially release both CuO and ZnO nanoparticles into the environment, posing environmental hazards. Specialized techniques and considerations are necessary for the recycling or recovery of CuO–ZnO nanoparticles, including nanoparticle separation and purification.
- **Environmental impacts in general:**
  The environmental effects of CuO/water nanofluid are mainly determined by the synthesis, dispersion, and utilization of CuO nanoparticles. Important factors to consider include the amount of energy used, the chemicals used, the amount of waste produced, and the potential environmental dangers related to CuO nanoparticles. The environmental effects of CuO–ZnO/water hybrid nanofluid are determined by the manufacturing, dispersion, and utilization of CuO and ZnO nanoparticles. Other factors to take into account include the cumulative impacts of CuO and ZnO nanoparticles on environmental performance, as well as the possibility of synergistic or antagonistic interactions.[4]

To summarize, the comparison of life cycle studies between CuO/water nanofluid and CuO–ZnO/water hybrid nanofluid entails assessing the environmental effects related to the extraction of raw materials, creation of nanoparticles, manufacturing processes, usage of the products, and disposal or recycling at the end of their life cycle. Although both nanofluid formulations have parallels in their life cycle stages, the presence of ZnO nanoparticles in the hybrid nanofluid brings about new factors to consider, such as nanoparticle compatibility, environmental dangers, and overall environmental performance.

10.4 SWOT (STRENGTHS, WEAKNESSES, OPPORTUNITIES, AND THREATS) OF NANOFLUIDS

10.4.1 Strengths

Nanofluids demonstrate a notable increase in thermal conductivity when compared to conventional heat transfer fluids. This characteristic allows for more effective heat exchange and enhanced energy efficiency in a wide range of industrial applications. Nanofluids and hybrid nanofluids provide a high degree of adaptability in their composition and use, making them suitable for a wide range of industries including electronics cooling, renewable energy systems, manufacturing processes, and biological applications. Novelty in nanotechnology enables the customization of nanofluid properties, such as particle size, shape, surface chemistry, and dispersion characteristics, to meet specific industrial needs and improve performance.[5]

Most importantly, nanofluids improve heat transfer rates, decrease system size and weight, boost equipment reliability, and prolong operating lifetimes, resulting in overall performance optimization in industrial processes.

10.4.2 Weaknesses

The exorbitant expenses associated with nanoparticles and manufacturing procedures pose a substantial obstacle to the broad acceptance of nanofluid technologies. This hinders the economic viability and scalability of these technologies, especially in cost-conscious industrial sectors. Nanofluids may encounter stability issues such as particle aggregation, sedimentation, and deterioration over time. These challenges can compromise the performance and reliability of industrial operations, particularly in extreme working circumstances.

The production of nanofluids is a complex process that involves the synthesis of nanoparticles, dispersion, and formulation of the fluids. It requires specialized equipment, experience, and quality control methods. These complexities might make it challenging to scale up production and achieve mass production.[6]

The presence of nanoparticles in nanofluids gives rise to issues regarding health and safety, specifically in relation to occupational exposure, inhalation risks, and environmental effects. As a result, it is necessary to conduct comprehensive risk assessments, ensure compliance with regulations, and implement protective measures in industrial settings.

10.4.3 Opportunities

Ongoing developments in the synthesis of nanoparticles, modification of their surfaces, and techniques for dispersing them present possibilities for enhancing the stability, performance, and cost-effectiveness of nanofluids. This drives innovation and broadens the range of potential applications. The increasing need for energy-efficient solutions, advanced materials, and sustainable technologies has led to a rise in demand for nanofluids in emerging fields such as advanced electronics cooling, renewable energy systems, and biomedical devices. This has resulted in market growth and diversification.

Collaborative partnerships between researchers, manufacturers, policymakers, and industry stakeholders promote the sharing of knowledge, transfer of technology, and development of markets, which speeds up the process of industrializing and commercializing nanofluid innovations. The sustainability focus prioritizes the development of eco-friendly nanofluid formulations, green synthesis processes, and recyclable materials, in line with global sustainability goals and regulatory requirements, to promote environmental responsibility.[7]

10.4.4 Threats

Nanofluid manufacturers have hurdles in complying with evolving regulatory frameworks, safety standards, and environmental restrictions. They must adhere to strict requirements and use risk management procedures to ensure compliance with regulations and gain market access. The market for nanofluid solutions faces significant challenges in terms of market penetration and adoption due to intense competition from alternative heat transfer fluids, conventional technologies, and emerging materials. To stay competitive, it is crucial to employ differentiation, value proposition, and market positioning strategies.

The presence of market volatility, economic swings, and geopolitical issues introduces uncertainty and risk into the nanofluid market, which in turn affects investment decisions, availability of funding, and market demand for nanofluid technology. Public perception has a crucial role in shaping consumer acceptance, regulatory scrutiny, and market demand for nanofluid goods. Misconceptions and concerns about the safety of nanoparticles, their environmental impact, and potential health risks can significantly impact the growth and adoption rates of these products.[8]

10.5 FUTURE PERSPECTIVES OF HYBRID NANOFLUIDS

The industrialization prospects for nanofluids and hybrid nanofluids are extremely promising, as they have the potential to bring about a revolutionary transformation in multiple industrial sectors. Nanofluids are expected to have a significant impact on improving heat transfer, energy efficiency, and performance in various applications, as breakthroughs in nanotechnology, materials science, and manufacturing processes progress (**Figure 10.2**). This analysis provides an in-depth examination

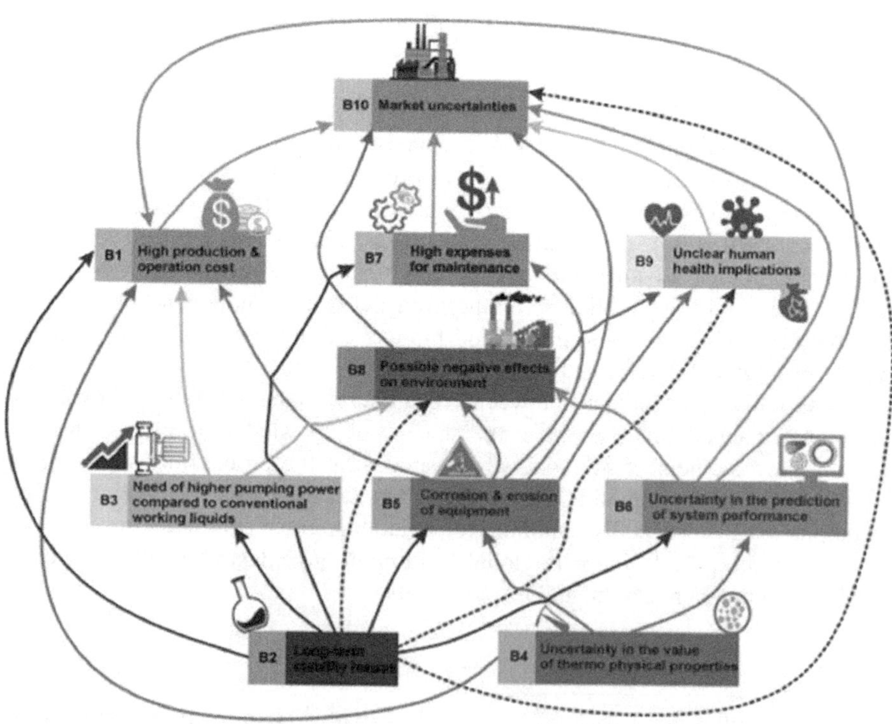

FIGURE 10.2 Prospects for the future of nanofluids.[9]

of the future prospects for nanofluids and hybrid nanofluids, specifically focusing on their potential for industrialization.

- **Improved thermal conductivity and increased energy efficiency:**
  Nanofluids possess enhanced thermal conductivity in comparison to conventional heat transfer fluids, hence facilitating more effective heat transfer in industrial applications such as cooling systems, heat exchangers, and thermal management devices. Nanofluids offer a promising solution for improving energy efficiency, lowering operating costs, and enhancing system performance, especially in high-temperature situations where traditional fluids have limitations.
- **Wide-ranging industrial applications:**
  The adaptability of nanofluids and hybrid nanofluids allows for a diverse array of industrial applications in several areas, including enhancing heat dissipation and lubrication, reducing tool wear, and minimizing processing times in manufacturing processes such as machining, metalworking, and plastic injection molding.

  Nanofluids are becoming more commonly used in thermal management solutions for cooling electronics, LED lighting systems, and managing the temperature of batteries. This is important since effective heat dissipation is crucial for the performance and reliability of these devices.

  Nanofluids are used in renewable energy systems, such as solar thermal collectors, geothermal heat pumps, and concentrated solar power plants, to improve heat transfer rates. This leads to higher energy capture efficiency and overall system output. Nanofluids are used in transportation for many purposes such as cooling automotive systems, lubricating aviation engines, and recovering heat. These applications help improve fuel efficiency, reduce emissions, and boost vehicle performance.

  In the biomedical and pharmaceutical sectors, nanofluids are used to achieve accurate temperature regulation in bioreactors, drug delivery systems, and medical imaging devices. This allows for progress in drug production, tissue engineering, and diagnostic technologies.
- **Customized nanoparticle design:**
  Advancements in nanoparticle engineering and surface modification techniques allow for the creation of nanoparticles that have customized properties and functionalities that are specifically optimized for particular industrial uses. Engineered nanoparticles that have precise control over their size, shape, surface chemistry, and ability to disperse provide improved stability, compatibility, and performance in nanofluid formulations. This drives innovation and allows for customization in industrial processes.
- **Nanofluids with several functions:**
  Hybrid nanofluids offer novel possibilities for improving performance and functionality in industrial applications by combining several types of nanoparticles or including additives into nanofluid formulations. Future research is to create hybrid nanofluids that have several functions and exhibit synergistic qualities. These properties include improved thermal conductivity, controlled rheology, resistance to corrosion, and antibacterial

activity. The goal is to design these nanofluids to meet specific needs in various industries.

Manufacturing solutions that can be easily expanded or adjusted to accommodate increased production demands. Scalable nanofluid manufacturing technologies and production methods must advance to be widely adopted in industry. Future trends encompass the advancement of continuous flow synthesis processes, scalable dispersion techniques, and modular production systems specifically designed for large-scale manufacture. The use of digitalization, automation, and process optimization technologies improves the efficiency, quality control, and scalability of nanofluid production, resulting in cost reduction and faster commercialization.

- **Factors related to the long-term viability and ecological impact:** With the growing significance of sustainability in industrial operations, upcoming advancements in nanofluid technologies are focusing on creating environmentally friendly formulations, utilizing renewable resources, and adopting sustainable production techniques. Green synthesis techniques, biodegradable nanoparticles, and recyclable nanofluid formulations reduce harm to the environment, enhance the efficient use of resources, and support sustainability goals, thereby promoting responsible manufacture of nanofluids.
- **Adherence to regulations and safety standards:** To assure the quality, reliability, and protection of human health, it is imperative to follow regulatory rules and safety requirements when industrializing nanofluids in the future. Regulatory bodies establish guidelines for the analysis of nanomaterials, evaluation of their harmful effects, and control of potential risks. These guidelines help manufacturers adhere to safety rules and encourage the responsible use of nanofluids in industrial applications.

10.6 SUMMARY

The utilization of nanofluids and hybrid nanofluids in industrial processes has significant potential to improve heat transfer, energy efficiency, and overall performance in different industries. However, it is crucial to tackle the problems described earlier about production, stability, cost, safety, environmental concerns, and performance optimization to fully commercialize and widely utilize them in industrial settings. Effective collaboration between researchers, engineers, and industry stakeholders is crucial to address these problems and fully realize the revolutionary advantages of nanofluid technologies in industrial applications.

REFERENCES

1. Kumar, L. H., Kazi, S. N., Masjuki, H. H., & Zubir, M. N. M. (2022). A review of recent advances in green nanofluids and their application in thermal systems. *Chemical Engineering Journal*, *429*, 132321. https://doi.org/10.1016/j.cej.2021.132321
2. Thakur, P., & Sonawane, S. S. (2019). Application of nanofluids in CO_2 capture and extraction from waste water. *Journal of Indian Association for Environmental Management (JIAEM)*, *39*(1–4), 4–8.

3. Thakur, P., Sonawane, S. S., Sonawane, S. H., & Bhanvase, B. A. (2020). Nanofluids-based delivery system, encapsulation of nanoparticles for stability to make stable nanofluids. In *Encapsulation of Active Molecules and Their Delivery System* (pp. 141–152). Elsevier.
4. Thakur, P. P., Khapane, T. S., & Sonawane, S. S. (2021). Comparative performance evaluation of fly ash-based hybrid nanofluids in microchannel-based direct absorption solar collector. *Journal of Thermal Analysis and Calorimetry, 143*, 1713–1726.
5. Thakur, P., Sonawane, S., Bhaisare, S., & Pandey, N. (2021). Enhancement of pool boiling performance using SWCNT based nanofluids: A sustainable method for the wastewater heat recovery. *Journal of Indian Association for Environmental Management (JIAEM), 41*(4), 7–18.
6. Sonawane, S. S., & Thakur, P. (2022). Current overview of nanofluid applications. In *Applications of Nanofluids in Chemical and Bio-Medical Process Industry* (pp. 1–26). Elsevier.
7. Sonawane, S. S., Thakur, P., & Chaudhary, R. G. (2022). Thermo-physical and optical properties of the nanofluids. In *Applications of Nanofluids in Chemical and Bio-medical Process Industry* (pp. 27–52). Elsevier.
8. Sonawane, S. S., Thakur, P. P., Malika, M., & Ali, H. M. (2023). Recent advances in the applications of green synthesized nanoparticle based nanofluids for the environmental remediation. *Current Pharmaceutical Biotechnology, 24*(1), 188–198.
9. Krishnan, S. S. J., Malika, M., Sharifpur, M., Sonawane, S. S., Mahian, O., & Meyer, J. P. (2023). Progress and challenges in nanofluids research. In *Nanofluid Applications for Advanced Thermal Solutions* (pp. 327–348). https://doi.org/10.1016/B978-0-443-15239-9.00012-6

Index

U

V

W

Z

For Product Safety Concerns and Information please contact our EU representative GPSR@taylorandfrancis.com Taylor & Francis Verlag GmbH, Kaufingerstraße 24, 80331 München, Germany

Batch number: 10397790

Printed by Printforce, the Netherlands